The Remarkable Reefs of Cuba

Hopeful Stories from the Ocean Doctor

David E. Guggenheim

Prometheus Books

Guilford, Connecticut

Prometheus Books

An imprint of Globe Pequot, the trade division of
The Rowman & Littlefield Publishing Group, Inc.
4501 Forbes Blvd., Ste. 200
Lanham, MD 20706
www.rowman.com

Distributed by NATIONAL BOOK NETWORK

British Library Cataloguing in Publication Information Available

Library of Congress Cataloging-in-Publication Data

Name: Guggenheim, David E., 1958–, author.
Title: The remarkable reefs of Cuba : hopeful stories from the ocean doctor / David E. Guggenheim.
Description: Lanham, MD : Prometheus, [2022] | Includes bibliographical references. | Summary: "The Remarkable Reefs of Cuba tells the story of the demise of the world's ocean ecosystems, the hard work of those desperately trying to save it, and an unexpected beacon of hope from Cuba—an island full of mystery and surprises"— Provided by publisher.
Identifiers: LCCN 2021057489 (print) | LCCN 2021057490 (ebook) | ISBN 9781633887800 (cloth) | ISBN 9781633887817 (epub)
Subjects: LCSH: Coral reef conservation—Cuba. | Coral reef restoration—Cuba. | Coral reef management—Cuba. | Marine biodiversity conservation—Cuba.
Classification: LCC QH77.C89 G84 2022 (print) | LCC QH77.C89 (ebook) | DDC 333.95/53153—dc23/eng/20220104
LC record available at https://lccn.loc.gov/2021057489
LC ebook record available at https://lccn.loc.gov/2021057490

♾️™ The paper used in this publication meets the minimum requirements of American National Standard for Information Sciences—Permanence of Paper for Printed Library Materials, ANSI/NISO Z39.48-1992.

All photos by the author unless otherwise noted.

This book is dedicated to Irene Hooper (1935–2014), founder and director of Seacamp Association (Big Pine Key, Florida), and Dra. María Elena Ibarra Martín (1932–2009), director of the Center for Marine Research at the University of Havana (Havana, Cuba). Each of these remarkable women, their desks just 125 miles apart on opposite sides of the Florida Straits, shared so much in common. With their selfless dedication to education and the oceans, they enriched the lives of thousands, engendering a love of and curiosity about the exquisite waters we share. I am privileged to have known, worked with, and had my life profoundly changed by each of them.

Plant a tree, have a child, and write a book.—José Martí (1853–1895),
Cuban poet, writer, philosopher, and nationalist leader

CONTENTS

PART TWO
ACCIDENT OF HISTORY

PART THREE
FORBIDDEN FRUIT, FORBIDDING POLITICS

PART FOUR
TIME TRAVEL

PART FIVE
SEÑOR, THE WORLD IS ABOUT TO CHANGE

PART SIX
LAND OF HOPE AND DREAMS

PREFACE

Tucked away in a hidden corner of Prague's old town lies the Church of St. Martin in the Wall. On Wednesdays and Thursdays, you'll find several dozen tourists milling about the entrance; a variety of languages and accents carry in the afternoon breeze. Once inside, the diverse group eagerly waits until the sublime strains of Mozart, Vivaldi, and Bach fill the modest church. As this rich musical tapestry unfolds, the barriers of place, race, and tongue are transcended by the universal language of music, uniting strangers in mutual love of a timeless and borderless art form.

A world away, biologists from the Caribbean watch the Gulf of Mexico swallow the sun at Guanahacabibes, Cuba's unpronounceable westernmost point. They, too, are united by a mutual love of a timeless art form: the green sea turtle. They have come together to help this endangered reptile, whose journeys encompass thousands of miles of open ocean, crossing the maritime borders of dozens of countries. Here on this small beach, Americans and Cubans side by side, and in those moments, the tortured political history between their countries is a distant and irrelevant absurdity. Bathed in the sun's fading golden light, the group silently awaits the first female turtle of the evening—having completed decades at sea to return for the first time to the very beach where she was born, feeling gravity's pull for the first time since she was a hatchling scampering for the surf, as she hauls herself up the beach to carve a hole with her back flippers and, seemingly in a trance, she lays her eggs. Afterward, as she swims into the black, she will unknowingly cross many political borders, focused only on her ancient will to survive.

At its essence, this book is a story of unthinkable environmental tragedy—the loss of one of our planet's most important and treasured ecosystems: coral reefs. But it is also the story of unexpected hope that offers a road map to guide our hand toward restoring the magic and beauty of an ecosystem so vital to the survival of our oceans, toward restoring the awe that has touched our hearts and sparked our imagination. Cuba's remarkable reefs offer us real stories of genuine hope. But it's impossible to tell this story without the context of the history of how we have managed to lose half of our coral reefs in the Caribbean. It's impossible to tell this story without telling the story of Cuba—its history, its culture, its people, and its struggles. It's impossible to tell this story without telling the inspirational story of collaboration of dedicated scientists from Cuba and the United States, working against a relentless tide of politics to learn from the sea and work to protect it. And it's impossible to tell this story without sharing my personal experiences, the victories and failures, from the perspective of an American with no Latin heritage, groping his way through unfamiliar territory for more than two decades, and the unanticipated reward of the profound warmth, welcoming, and friendship of the Cuban people.

To convey so much in one book, my pen and my scalpel have necessarily sat side by side on my desk. In the end, this is far from a textbook on coral reef science and conservation. Nor is it anywhere near a comprehensive history of Cuba or a treatise on Cuba–U.S. relations and regulations. No doubt there are glaring omissions from each of these topics. Rather, it is an account not only about science and history, but also a collection of stories of real people, real challenges, and real work on the ground—or in the water. It is a journey through what I consider simultaneously the most rewarding work of my career and the most frustrating, a roller coaster screaming along the tracks from the heights of euphoria to the depths of despair.

A few notes about the contents: The book contains dozens of interviews of friends and colleagues, Cuban, American, Bajan, and Russian. For some, it also includes their eloquent words while guest lecturing to my graduate students at Johns Hopkins University. For clarification, the title of this book refers to "the Ocean Doctor," a moniker given me by my daughter, Anna many years ago. In the body of the book, I sometimes refer to Ocean Doctor, which is also the name of the nonprofit organization I run. In some cases, I have changed the names of individuals to avoid bringing them unwanted attention or risk. I have worked under the umbrella of several organizations over the 20-year period covered by this book. For simplicity's sake, I have not always specified which organization I was with at the time, opting instead for pronouns like "we."

Telling "stories of hope," as the title promises, requires understanding why hope is needed in the first place. I therefore must drag my readers through stories that are much less hopeful—some downright depressing—in order to set the stage of why Cuba's coral reefs are truly remarkable and why they indeed offer hope. So when it seems that all hope may be lost, hang in there. I promise to leave you on a high note.

Though I tell this story from my own perspective, I do not want to give the impression I am by any means alone in collaborative marine science and conservation work in Cuba. I am joined by other capable, dedicated, and respected organizations from the United States and around the world. In the same vein, this book focuses on collaboration between the United States and Cuba, but my intent is that the material also connects with non-Americans and non-Cubans, as the story of coral reefs is a global one. But, hey, I'm an American. . . . Also, this book certainly involves science, but it is not written for scientists. My goal has been to ensure that the material is not only understandable by nonscientists, but also that they will not find such material dry and boring.

The coral reefs I grew up with—ecosystems of indescribable beauty and sanctuary that literally changed the path of my life at the age of 15—are rubble today. Ninety percent of the reefs I cherished in the Florida Keys have perished. But when I nearly gave up hope that coral reefs would survive to the end of the 21st century, Cuba came to my emotional rescue. I beheld thriving coral reefs even healthier than those I remember as a teenager. Along with the science, history, and politics framing this story, I hope I also am able to convey the emotion of this journey of hope. After more than 100 trips to Cuba, I find myself more and more tongue-tied to describe the magic of this island that draws me back again and again. It's been described as a paradise. I resist using this word—it has been commandeered by travel marketers and overly boastful coastal resorts around the world. But the word certainly fits. Cuba is indeed a paradise, not only of incredible, stirring natural beauty, but also of an incredible, resilient, and gentle people, whose warm hearts have welcomed me again and again, and where I have forged some of the most meaningful friendships of my life. And now, at a time when all may seem lost, it is a paradise that offers us the most precious gift: hope.

Unlike the mother sea turtle plying the ocean's waters, political borders are all too visible for our species, and working across those borders is frequently complicated, unpredictable, and often agonizingly slow. The hallmarks of international work are patience, curiosity, empathy, and humility. It's not enough to know the science and speak the language. It's critical to understand the culture and politics, which are inescapably intertwined with conservation. The oceans once stood as a forbidding, impenetrable divide that held civilizations apart for centuries. Today, we are united by our growing concern for the health of the oceans. And, like the beach at Guanahacabibes, it is more important than ever to find the common bond of conservation that will unite us in partnerships that will endure across languages, borders, and politics.

PROLOGUE

My tired eyes panned frantically but in vain for even the slightest evidence that I was in my lane—or any lane—as we sped toward an awaiting vessel in a tiny coastal village I had never heard of. Our deadline was nightfall, which had been ominously enveloping us. The Cuban Coast Guard officials at our destination would soon pack up and return home to their families, decreeing it was too late for the *Reina* to leave port.

We were hours late, stymied by typical Cuban obstacles, including trying to rent a pair of reliable cars for the journey from Havana, an ordeal that took fully half a day, punctuated by bursts of exhaustive negotiations, fists full of cash, and crosstown dashes to promised vehicles that didn't exist. We were also slowed by the inevitable breakdown of one of our "reliable" vehicles along the highway. Thanks to a pair of tie wraps and Cuban-inspired ingenuity—ironically performed by American NPR reporter and *Washington Post* writer Nick Miroff—we were able to jury-rig a repair.

The six-hour drive east along Cuba's major axis became increasingly harrowing with every minute as a determined sun slid steadily into the mountains behind us. My grip on the wheel tightened until my hands throbbed. In Cuba, darkness transforms a drive in the countryside from a mildly challenging sojourn into a nail-biting adventure, especially on this final stretch of highway leading into Ciego de Ávila Province.

As the pale gloaming gave way to darkness, oncoming headlights scattered across our filthy windshield, causing momentary night blindness, frustrating whatever hope I had of being sure I was in my lane. Such certainty would have provided a morsel of desperately needed comfort, though far from a guarantee of safety.

This stretch of the *autopista* had three narrow lanes, impossible to discern in the darkness. "Which direction is the middle lane?" I asked my Cuban friend and colleague, Dr. Fabián Pina, who lived somewhere beyond the far end of this highway. His matter-of-fact response confused and terrified me.

"It's both directions. It's a passing lane," he replied.

I protested, "You've got to be kidding!"

Eighteen-wheelers with blinding headlights barreled toward us in the center lane, seemingly inches away, their explosive wake—together with my own instinctive reaction to steer away—lurching our small station wagon toward the dirt shoulder, where invariably our headlights would reveal the rapidly approaching rear end of a horse and the unlighted cart it was pulling. So I'd veer to the left and hold my breath, hoping we weren't coming up on a slow-moving vintage 1950s Chevy without working taillights.

This maddening dance down the highway continued into the early evening as the *autopista* slowly gave way to the narrow two-lane secondary roads that wound through acres upon acres of sugarcane. When I finally released my death grip at a small-town service station, Nick observed the bizarre mix of horse-drawn buggies, bicycles, decrepit old American cars, rusted Russian Ladas and Moskvitches, and modern Asian and European vehicles we had encountered. "It's like the whole history of human transportation on one road," he said.

This moment—and countless more throughout each of more than 100 visits to this unusual island—seemed to be the realization of a science fiction book I had enjoyed as a teenager. In Gordon Dickson's *Time Storm*, swirling storms in time crisscross the planet, leaving in their wake a world divided into a patchwork of different time periods. The story's protagonist must journey through the often-bizarre juxtaposition of different time periods in close proximity.

And so also to the outsider does Cuba seem to be caught in its own time storm, with relics of a century past coexistent with early 21st-century modernities: horse-drawn buggies stopped at a traffic signal alongside modern Hyundai sedans; 60-year-old rotary phones still in regular use alongside shiny new iPhones. It is disorienting to travel through an island where so much is still frozen in time, almost as it was more than a century ago, with more than half a century of isolation and crushing economic embargo from its nearest neighbor, the United States.

When I first set foot on this exotic island, little did I know what lay in store over the next two decades. I underestimated everything, including Cuba's relentless gravitational attraction that would pull me back time after time, even after I had given up hope of succeeding there. Nor did I appreciate what difference I could possibly make, a gringo born in Philadelphia without a drop of Latin blood in my veins and little memory of the Spanish I had struggled with during my undergraduate studies decades earlier. But I would soon discover—and treasure—my newfound Latin soul.

Neither had I anticipated that I would be swept away within my own "time storm." As we resumed our mad dash to the tiny coastal village of Jucaro before the Coast Guard officials left, the awaiting *Reina*, a converted fishing boat, was to be my shiny silver DeLorean, destined to take me on a journey through time that I can still barely believe, one that would bring me face-to-face with vibrant, healthy coral reefs, reefs I thought I would never lay eyes on again, reefs that were the inspiration of my career in marine science and conservation, reefs that would soon succumb to a tidal wave of humanity invading Florida and lethally warming seas. And so would go much of the Caribbean reefs as well. The coral reefs my colleagues and I had so cherished in the seventies were 80 to 90 percent dead. And in the Caribbean, the figure was roughly 50 percent.

At a time when I had come close to giving up hope for corals in the same way that many conservation groups abandoned the Caribbean as a lost cause, Cuba suddenly offered unexpected hope. Just like the dazzling carpet of schoolmasters, grunts, tang, and parrotfish spilling over the mustard walls of Florida's Looe Key so transfixed me that summer in 1974 as a young teenager, so did my eyes widen and my heart race with joyful disbelief as I slipped below the surface from the stern of *Reina* into an underwater paradise as frozen in time as the rest of Cuba. Little had I suspected that an old, converted Cuban fishing vessel was the time machine that I and many others had fantasized about, to take us back in time to the days of flourishing reefs and with that journey, perhaps a second chance to do things right.

I beheld magnificent stands of healthy elkhorn coral, teeming with colorful grunts, snappers, and angelfish. I came face-to-face with Goliath groupers, a critically endangered species, more than triple my weight. I found myself surrounded by dozens of healthy Caribbean reef sharks, silky sharks, tarpon, and myriads of other sprightly fish and corals, all the while seeing no evidence of the decay and disease the rest of the Caribbean had suffered over the past half century.

The experience was so powerful it would inspire some of the most important work of my career. But it would also leave me humble and speechless, groping for adjectives that don't exist, reminiscent of explorer Meriwether Lewis's emotional encounter with the Missouri Falls in 1805 during the famed Lewis and Clark expedition, excitedly describing the falls in his journal as "the grandest sight I ever beheld." But after rereading his words, he scrawled his exasperation and despondence in a rambling apology to the reader, punishing himself for offering words that so inadequately conveyed the grandeur of the wondrous sight he beheld. I would spend years searching for the right words to describe coral reefs, but, as with Meriwether Lewis, they continue to elude me.

PART ONE
CORAL'S LAST STAND

Success is survival.

—Leonard Cohen

CHAPTER ONE

SPINELESS

The last fallen mahogany would lie perceptibly on the landscape, and the last black rhino would be obvious in its loneliness, but a marine species may disappear beneath the waves unobserved and the sea would seem to roll on the same as always.

—G. Carleton Ray

The Shark Pit

The fourth graders—newly arrived and attending their first class—were unmistakably nervous, tightly clutching their towels, masks, fins, and snorkels that morning in late 1974. They had just been told that they would be swimming with sharks in a few minutes. Freshly arrived from a snow-swept Midwestern state, they stood pale and unsettled in the morning sun, smeared with fragrant suntan lotion as they quietly listened to Bob Beech, their instructor at Seacamp/Newfound Harbor Marine Institute, a nonprofit marine science camp and education institution in the Florida Keys.

Founded in 1966, Seacamp was among the first such marine education centers dedicated to teaching young students marine science and doing so with "experiential education," that is, tossing kids into the water to behold coral reefs or leading them on a tromp through the muck of a subtropical marsh, all the while enduring the stings of *Cassiopeia* jellyfish

3

from below and swarms of unrelenting mosquitoes from above. Experiential education means experiencing Mother Nature on her own terms, typically euphoric, occasionally terrifying, at times maddeningly uncomfortable, but always profoundly illuminating. Such experiences could be truly transformational, taking kids out of their day-to-day lives—often devoid of nature—and plunging them into a thriving natural world they had never before experienced. For many, like myself, it led to a lifelong love affair with the coral reef ecosystems of the Florida Keys and the life, color, and magic they harbored.

I discovered Seacamp perusing *Boys' Life* magazine, where I spotted a small ad in the summer camp classifieds with the tantalizing tagline, "For all the sea has to teach us . . . and all the fun in learning it." So, on June 24, 1974, I boarded Eastern Airlines flight 35 in Philadelphia and sat myself in seat 12A. The Boeing 727 rumbled down the runway close to our scheduled 9:00 a.m. departure, and three magical hours later, a 15-year-old teenager from Philly found himself in Miami, Florida, eager with anticipation of catching his first glimpse of the Florida Keys . . . wherever *they* were. I didn't know. Donna Coffin, the thoughtful and kind redheaded counselor who met me at the Eastern Airlines gate, kindly drew a map on a napkin. I was astonished—and excited—that Florida continued south along a chain of islands that I never knew existed. I would eventually spend part of 11 summers at Seacamp, three as a camper, eight as an instructor.

The fourth graders were assembled adjacent to Seacamp's dining hall where there was a small man-made saltwater pond the size of a large circular swimming pool. From the center of its dark green waters emerged a white sign on which was painted the menacing moniker: "SHARK PIT," sporting a crudely painted silhouette of a shark. "We'd sit on the back seawall and catch sharks and throw them into the Shark Pit . . . small sharks," recalls Bob. Lemon, nurse, and bonnethead sharks circled the pond. "Even those little sharks, they would cut the surface with their dorsal fin," leaving a small wake behind them, just like their fully grown counterparts, "so getting fourth graders to go in and snorkel to see them firsthand was a psychological challenge," Bob continues.

Bob was my marine ecology instructor at Seacamp, and a few years later, my colleague as I rose from the ranks of camper to become an instructor. Bob is as gregarious today as he was then. In the seventies he stood tall and tanned, his sun-drench curls streaked with highlights of blond beneath his ever-present white tennis visor and his beard framing a welcoming and often mischievous smile. In the mid-seventies, we were now squarely in the post-hippie era, but Bob still harbored all the spirit, energy, curiosity, and independent thinking of that period in his blood. We'd have long talks about life, often venturing into philosophy. But deep within Bob was an indomitable entrepreneurial spirit, and that was the force that took him into a career rich with startups and corporate leadership in biotech and other disciplines. He's far from the Keys now— in Cincinnati, with his wife Christine, another Seacamper. Despite the passage of more than four decades, this particular day, with these fourth graders, still makes him wince.

"So, I had sort of developed this little schtick. I'd get down there and I would be the first in, and of course, this was after a little talk to them saying these are little tiny sharks, they can't harm you and it'll be fun." For this group, it was going to take a bit more convincing. The students stood their ground, no doubt the theatrical release of *Jaws* a few months earlier fresh in their minds. "Usually you could tell who was going to be the brave kid. It was kind of like 'me first' as they were getting closer to me."

The dark green, algae-rich waters of the Shark Pit made it challenging to see the sharks on a good day, and if the muddy bottom was kicked up, you could barely see your hand in front of your face. As he had done with many classes before, Bob instructed the students to follow his lead and not wear their fins. He left his fins at water's edge and proceeded barefoot, mask and snorkel in hand.

The fourth graders stood silently and watched in a mix of fascination and horror as Bob edged closer to the water. "So, I went through my normal spiel . . . turned around, and I went to walk into the water. I walked . . . two or three steps and . . . stepped right on it—a fucking *Diadema* that somebody had thrown in there!" A *Diadema*, technically *Diadema antillarum*, is a jet-black, spherical sea urchin covered with thin, razor-sharp spines that can grow to more than a foot long. They're not poisonous and are generally harmless to humans—unless you don't have the good sense to not step on one with all of your body weight. Their spines can easily penetrate flesh and lodge themselves deep below the surface of the skin, inflicting the exquisite pain one might feel by stepping on a dozen thick hypodermic needles. However, while you can simply pull a hypodermic needle out, *Diadema* spines are practically impossible to remove. They're brittle and break off, making extraction a nearly futile exercise, so victims must painfully wait the days or weeks until the spines gradually dissolve. Many of the Seacamp campers and instructors, myself included, bore the badges of *Diadema* encounters, little black specks peeking from beneath skin.

"I impaled myself, and I remember screaming and falling backward," Bob continues. The students, unaware of the unseen *Diadema*, had come to the logical conclusion that any fourth grader would: A shark had decided to make Bob its breakfast. "And as I'm falling backward, it's like these little smoke trails of kids were going up the property. . . . They were hightailing it out of there. They thought for sure that my leg had been taken off."

It took some time for staff to locate the now even-paler, terrified students and convince them to return to the dreaded Shark Pit. "Then of course we herd them back and I'm there and I've got these spines stuck in the bottom of my foot," Bob recalls. Being the good teacher he was, he desperately tried to improvise, trying to take mayhem and transform it into a teaching moment. "The *Diadema* was missing a third of its spines . . . and I was trying to educate them about it." It was all but futile—the students could only stare at this man with black spines in his foot who had been shrieking just a few minutes earlier. "I don't even remember if they ever got in the water."

But Bob's challenges with the fourth graders didn't end there. Several days later he was assigned to teach the same group of students, this time to take them out by boat so, for the first time in their young lives, they'd be able to see corals. Bob boarded the students onto a flattop boat and took them a few miles to a small group of nearshore coral heads. "As soon as they got in the water they looked down; what did they see? All these *Diademas!*" Bob remembers. They rocketed out of the water. "They were scared to death, so the whole thing was just a mess." Bob did not let it go. "I paraded around the property after digging those spines out trying to find out who the hell had thrown a *Diadema* in there. I still have memories of pain. When you get those things stuck in you far enough, it's a major ordeal. I think I hobbled around for several weeks."

Diadema normally don't hang out in Shark Pits—they relish the clear, shallow waters of coral reefs and crowd around their edges, sometimes squeezing into crevices hidden beneath their ledges. They were loathed by snorkelers and divers who carelessly brushed up against them . . . or occasionally stepped on them. But little did we know, nor could we imagine, that in a few short years *Diadema* would go from reviled to revered.

Halo Makers

"The reefs were black!" Initially I didn't quite understand what friend and colleague Dr. Don Levitan meant. Did he mean that the reefs were covered in oil or some other contaminant? Had they died? "You couldn't even get into the water because the *Diadema* were so thick." With lingering incredulity and marvel from what he observed as a budding scientist in 1982, Don described how the density of *Diadema* along the coast of St. John in the U.S. Virgin Islands was 15 to 30 urchins for every square meter. In less technical terms, that's wall-to-wall urchins. Don continued, "The only way to get in the water was to jump off a boat or a dock . . . you couldn't walk into the water."

Don is another ex-Seacamper. With a mop of unkempt dark brown hair and a bushy beard, his typical Seacamp attire—much like the rest of us—was frayed jeans cutoffs and a soiled T-shirt. Don had boundless energy and intensity, tempered by a warmth and effusive sense of humor that endeared him to everyone. He taught scuba diving and several marine science classes, and I served as his scuba class assistant on a number of occasions. Later, Don steered his career in a straight line toward academia, landing him in graduate studies at the University of Delaware and eventually at Florida State University as a professor of biology.

"It's a curious story because I was all set to go to graduate school to work on corals. I got this job teaching a School for Field Studies [a study-abroad program where students participate in research projects] course down in the Virgin Islands."

But once Don got to St. John, his focus on corals became secondary to the incredible masses of *Diadema* he beheld. "It occurred to me, how can these urchins persist at such high population densities?"

To the casual snorkeler or scuba diver, *Diadema* appear to be immobile. It takes time-lapse photography to catch them in the act: They constantly graze algae—their favorite delicacy—that's growing on the reef itself. At night they'll even sneak away from the reef for dinner, making their way around the reef to munch on the algae. It grows relentlessly upon the sandy substrate, and the *Diadema*, like little lawn mowers, leave swaths of clean sand behind them. Don recalls the results of hordes of *Diadema* at work in 1982: "If you looked at the substrate it was about as polished white as you can imagine."

Healthy coral reefs have a so-called halo around them, a pronounced perimeter of bare sand around the coral formations, devoid of algae thanks to algae-lovers like urchins and fish. They also keep the area clear of sea-grasses, like turtle grass. These halos are so reflective that they're visible from space. So, *Diadema*—eagerly fattening themselves up on a daily crop of algae—aren't just self-serving. They're halo makers, and halos matter. Don points out that by cleaning the sand, each *Diadema* "prepares the substrate for coral settlement." If you're a baby coral, you need that halo. Corals—whose larvae are free-floating, or planktonic, can travel great distances on ocean currents. If you're a lucky larva, you'll settle into the clean sands of these halos and get to grow up to be a big coral, unfettered by competition from other species—especially algae. In a battle royale between slow-growing corals and fast-growing algae, it's the algae that handily win in the first round. Unchecked, algae can easily smother and kill a coral reef. Grazing by the *Diadema* "also reduces . . . nasty chemicals that algae can produce that can inhibit coral growth," Don observes. But it doesn't end there. It turns out that *Diadema* are also little coral reef paramedics. Don explains: "Urchins would graze *over* the corals and . . . if a coral was wounded, they would clean it. I think that also potentially helped against disease. . . ."

Don's fascination with *Diadema* grew. "So, I switched my dissertation from thinking about corals to thinking about these urchins." The timing of his decision was uncanny. Six months later, off the coast of Panama, something strange began to spread its way through the Caribbean, and it was happening at an alarming rate: *Diadema* were dying and dying en masse. A few months later, the *Diadema* had been virtually eliminated in the Caribbean. According to marine scientist and coral reef expert Dr. Nancy Knowlton at Smithsonian Institution, this event is still considered "the most extensive mass mortality ever reported for a marine organism." The *Diadema* were gone, and to this day, we don't know why.

Finding Ground Zero

Surrounded by lush mangroves and cradled by the warm Caribbean Sea, Panama's Punta Galeta is home to the Galeta Point Marine Laboratory, part of Smithsonian's Tropical Research Institute. Its striking natural beauty and rich wildlife stand in stark contrast to its dramatic backdrop a mile to the west: an unending parade of enormous ships from around the world. Punta Galeta marks the eastern edge of the Caribbean entrance to the Panama Canal.

It is in the waters of Punta Galeta that scientists first noted something odd about *Diadema* populations. They had good baseline numbers—the waters were adjacent to the research station. If you were a *Diadema* or any other marine creature, you could pretty much bet the bank that a scientist would eventually drift by with an underwater slate to count you. So, when Dr. Harilaos Lessios and his colleagues from the institute began to see *Diadema* vanishing, they could readily pull out the data from prior years and run the calculations to confirm the impossibility of what their eyes were seeing. Dead and dying *Diadema* lay everywhere. Strangely, they observed that only the *Diadema* and no other species were affected.

Over the course of just one year, the local *Diadema* population went into free fall, bottoming out at less than 1/100th of a percent of their former density (0.00357% to be exact), dropping from 14,000 animals per hectare in June 1982 to 0.5 animals per hectare in May 1983.

Lessios's team made the first detailed observations of the gruesome way that *Diadema* were perishing. The first symptom was a loss of the spines' ectoderm, the outer, protective covering. Without it, sediment began to accumulate on the spines. Soon, they began losing their distinctive black pigmentation. Their spines became brittle, many beginning to break off. Finally, the tiny hydraulic tube feet they use to cling to the bottom became flaccid and dysfunctional—they couldn't move, and even worse, they could no longer hold fast on the bottom. Unable to hold on, they were powerless against the surging waters, rolling helplessly along the bottom, losing more and more of their protective spines until, recognizing their vulnerability, a variety of fish went in for the kill, eagerly feeding on this new delicacy in their diet.

Lessios referred to it as a "mass mortality" event, but just how massive was it, he wondered. He frantically began contacting his colleagues

9

around the Caribbean, among them Nancy Knowlton. She and others began to confirm that Punta Galeta represented ground zero and that a dramatically larger event was already under way. It soon became clear that the phenomenon was spreading at lightning speed to the far reaches of the Caribbean Basin and even into the subtropical climes of the Gulf of Mexico and the Florida Keys.

"Reef scientists scattered across the region (in the days before email) so that they would be ready for it should the mortality spread. And spread it did ... ," recalls Knowlton. Initially it spread relatively slowly, "reaching Costa Rica to the west and Columbia to the east. . . . But then the spread accelerated, all the way to Bermuda." Populations were decimated to 2 to 7 percent of their former levels, sometimes within just days of the onset of the symptoms, decimating reefs that, as Don Levitan had observed, "used to be black with urchins." Now, Nancy recounted that, incredibly, one could swim for an hour without seeing a single *Diadema*.

Scientists were caught off guard by the speed and extent of this mortality and scrambled to understand the cause. In the years since, most scientists have settled on the same conclusion: "The speed with which this mortality occurred is part of the reason why, to this day, we still do not know the agent responsible, although the pattern of spread (largely following currents) and specificity (no other urchin was affected) make a pathogen a near certainty," concludes Knowlton. What that pathogen was remained a mystery.

Dr. James A. Bohnsack, recently retired as division director of the National Oceanic and Atmospheric Administration's (NOAA's) Southeast Fisheries Science Center in Miami, shed light on where much of the scientific community believes the mysterious pathogen originated and what it might have been. He described the theory in thoughtful and fascinating detail, as I remember he did when I took his classes in the seventies. Jim is yet another member of the Seacamp diaspora who launched his career on that small corner of Big Pine Key. More than my instructor, Jim was also my mentor. I was overwhelmed with pride when Jim asked me to be his lab assistant, cleaning test tubes, scraping fish guts off the floor, splashing coats of noxious epoxy paint onto drying racks—I was in scientist-wannabe heaven.

Jim has focused his career on fish—there's even a fish-counting method named after him. But far more than a fish biologist, Jim understands how all the pieces fit together to make a healthy coral reef ecosystem work and knows how to make it understandable for the rest of us. Over the years, my colleagues and I marveled at how Jim survived working for a federal agency. He'd probably be the last person you'd pick out of the room at a symposium to be a high-level NOAA employee: Hawaiian shirt, curled mustache, and an ever-present infectious laugh. Most astonishing is that in all the years he served with a federal agency, Jim never minced his words, nor did he spit jargon at his audience or colleagues. He has always been honest, straightforward, and an independent thinker, unafraid to go off-script. He always delivers the unvarnished facts, unburdened of politics and always with irreverent humor. For the life of me I don't know how he never got fired for such sins. Probably because he's so damn capable . . . and lovable.

Jim explained that most scientists speculate that the mystery of the "Great *Diadema* Die-Off" leads to the Panama Canal. There, hidden deep in the ballast water of a ship transiting to the Caribbean from the Pacific, lurked an unknown pathogen. Ballast water is pumped into special tanks and cargo holds of ships, used to provide stability and maneuverability. Once a ship unloads its cargo, ballast water compensates for the lost weight, allowing the ship to ride lower in the water, increasing its stability. Ships entering the Pacific side of the Panama Canal take on water to maintain stability against high winds in narrow passages, and then dump the bilge water once they emerge on the Caribbean side. Ships also periodically flush their bilges to keep marine organisms from growing and fouling tanks and pumps.

According to its 2009 report, World Wildlife Fund International estimates that at any given moment, 7,000 marine organisms are crisscrossing the globe, tucked away out of sight in the ballast water tanks of ships, many able to survive journeys of thousands of miles. If released, some can be invasive, pushing out local flora and fauna, leading to "irreversible ecological change and economic loss."

There are countless examples of such invasions by ballast water, such as the Chinese mitten crab, now found in the North Sea, the Baltic Sea, and the Atlantic and Pacific Coasts of North America. These crabs burrow into the sediment and have caused serious erosion of riverbanks as well as clogging of municipal water systems. The North American comb jellyfish helped decimate anchovy and sprat stocks in the Black Sea in the late 1980s. They have continued to spread to the Caspian, North, and Baltic Seas. In the waters of the Panama Canal itself is an unlikely resident: the Iraqi crab, which had only been reported previously inhabiting the Shatt Al-Arab River system at the confluence of the Tigris and Euphrates Rivers.

Perhaps even more chilling—especially considering the plight of the *Diadema*—ballast tanks can harbor far more than fish, jellyfish, and crabs. Trillions of microorganisms—including bacteria and viruses—may accumulate in a single tank of a single ship.

For decades, scientists have been locked in a heated battle over the possibility of a sea-level Panama Canal that would eliminate the lock systems and directly connect the Pacific and Caribbean. Of great interest to shipping and commerce, Congress began hearings on the topic in the sixties.

Such debates typically focused on species like fish, crabs, and clams, and it was thought highly unlikely that any of these species trying to invade the Caribbean would be successful, as the local critters had a head start of many millennia to adapt to those waters. The newcomers just wouldn't be able to compete.

Ballast water occasionally made its way into those discussions. At one such hearing in 1971, a scientific paper published a few months prior by Ira Rubinoff was entered into the congressional record as written testimony and illustrated the thinking of the time, namely, that the possibility of a ballast water invasion was extremely unlikely. In the paper, Rubinoff explained that "the environment in most ballast tanks is remarkably inhospitable . . . particularly for the relatively planktonic organisms that are most likely to be taken into ballast systems. Anti-corrosion paints that are used to protect these tanks are extremely toxic, and a few minutes contact with them is sufficient to kill most marine organisms." In the same paper, he may have sensed that the future would be different. He acknowledged that the next generation of tankers, equipped with tanks made of stainless steel, would help ensure "the successful carrying of marine organisms from ocean to ocean." Meanwhile, the presence of pathogens in ballast water— bacteria and viruses—was typically just a footnote in those discussions.

Jim Bohnsack pointed out that regulations were established restricting bilge flushing or pumping in the Panama Canal. "They stopped the ships from 'ballasting' for environmental reasons; they didn't want people putting this saltwater stuff in this freshwater lake."

Jim was referring to Lake Gatún, the world's largest man-made freshwater lake created in 1914 as part of the canal system. It sits between the Panama Canal's two sets of lock systems—one on the Pacific side and one on the Caribbean side. The lock system is slow and inconvenient for crossing ships. The average crossing time is 8 to 10 hours. But the presence of an enormous freshwater lake in the middle of the transit is seen as an environmental plus—if you're not the lake itself, that is. The fresh water serves as a buffer, so organisms hitching a ride on the hull of a ship cannot survive the hours immersed in fresh water that they're subjected to (though it does nothing to affect those organisms living inside the ship in the ballast water). Ironically, the well-intentioned decision to protect Lake Gatún's environment from saltwater ballast water dumping has likely come at the environmental expense of the Caribbean Sea.

"This is a theory now," Jim qualified his remarks. "What happens is there's a whole series of bacteria that forms cysts when they go anoxic— they lose oxygen and they make a cyst." A microbial cyst is a resting or dormant stage of a microorganism, usually a bacterium. The cyst is something like a suspended animation chamber—it helps the organism to survive when environmental conditions are unfavorable, like in the ballast water of a ship when there's little oxygen. Once things improve, the bacteria let loose.

"Because they couldn't flush out their tanks, when they got from the Pacific to the Atlantic, they released [their ballast water]," and the bacterial cysts within it, which "got loose there and killed the *Diadema*. It turns out echinoderms are notoriously vulnerable to this class of bacteria."

The explanation is plausible and logical, but unfortunately, we may never know if it's true. Not only did the scientific community fail to diagnose the disease at the time, but they didn't properly preserve samples so they could be analyzed at a time in the future when it might be possible to divine the answer. If only there were, tucked away in the back of someone's freezer somewhere, an afflicted *Diadema* from 1983, then modern, cutting-edge technology, unimagined in the early eighties, could be put to the task. Jim recounts, "Unfortunately, they didn't freeze it; they used formaldehyde to preserve everything, which destroyed any bacteria material."

We may never know what killed the *Diadema*, but decades later we know that they haven't come back. In a 2014 scientific paper published in *Oecologia*, Don Levitan concludes, "We find no evidence to support the notion that this historically dominant species will rebound from this temporally brief, but spatially widespread, perturbation." As a fellow scientist who has authored scientific papers, I'll cut Don some slack. Dispassion is important when putting pen to paper. But calling the *Diadema* die-off a "perturbation" is reminiscent of the *Challenger* explosion being referred to as a "malfunction." The *Diadema* die-off was a horror show. But as bad as it was, it triggered something far worse.

Death Blow

Scientists won't always state it conclusively in scientific papers, but over a beer most would agree: the "Great *Diadema* Die-Off," as it became known, marked the tipping point for many Caribbean corals, a death spiral that continues today.

Nancy Knowlton observes that "the consistency of ecological responses around the region leaves little doubt that [*Diadema*] was a keystone species at the time." If you're a keystone species, you're kind of a big deal. A keystone species is one on which other species in an ecosystem are highly dependent. Removing that species invariably leads to major changes in the ecosystem. If it hadn't been recognized before the die-off, it was now clear as day: A *Diadema*-less Caribbean would be a much different place. *Diadema* weren't there to mow the lawn anymore and no one else stepped up to do it. So, in short order, reefs were overgrown by algae. The widely used colloquial expression of this phenomenon came from the recently released (at that time) film *Ghostbusters*: The reefs had been slimed. The term wasn't far off. Some of the sticky, filamentous algae resembled the "ectoplasm" that covered Bill Murray's face.

Dr. Patricia González, former director of the University of Havana's Centro de Investigaciones Marinas, Center for Marine Research (CIM), whose passion is coral reefs, expressed her angst that once a reef has been slimed, it's very hard for it to recover. Like Don Levitan, she observes that algae not only covers the reefs but also the sand and substrate around them. Young coral larvae can't get a foothold. "So even if the reproduction process is successful, in the end the larvae can't find a good substrate upon which to settle." So the oceans are not only losing their live, mature corals, they're also losing the capacity to make new corals.

Is the story of Caribbean corals this simple? No *Diadema*—no coral reefs? Hardly. Corals are exquisitely sensitive to all manner of stresses. The problem is that we've been unleashing those stresses for decades. The *Diadema* die-off was, as Don Levitan put it, simply the "final death blow" for many coral reefs that were already experiencing more stress than we had realized, even back in the seventies and prior. "That's been my story going through the Caribbean now. Each place, something different comes and does the final death blow to that reef." Sometimes it was the loss of *Diadema*, sometimes a bleaching event, sometimes a powerful hurricane. "So, the stresses are pretty much the same, but the thing that causes the death blow varies from site to site."

There have been a lot of "death blows" for Caribbean corals. Roughly half of the Caribbean corals have died since 1970 and the prognosis for them surviving the balance of this century is chilling.

An Island of Hope?

As Caribbean coral reefs were unraveling, in the middle of the mayhem lay an island where little was known about the health of its coral reefs. This was especially strange as it is the largest island in the Caribbean. But this island was different. It was politically isolated, and contact with its scientific community was exceptionally rare. Over the years, Cuba has harbored many secrets, but lying at the confluence of the Atlantic, the Gulf of Mexico, and the Caribbean, the secrets it held beneath its waters were especially intriguing.

What was little known at the time outside Cuba is that it had not escaped the *Diadema* debacle. There, too, the *Diadema* populations plummeted. However, unlike nearly all its Caribbean neighbors, its corals remained remarkably healthy, almost as if nothing had happened. Cuba's marine waters continue to flourish today—a "living time machine" that has somehow endured the four decades after the loss of *Diadema*, many areas nearly indistinguishable from a time when the decades-old Chevys, Fords, and Edsels that now roam Cuba's streets sparkled as they rolled off the assembly line. The question I'm most often asked is, of course, "Why?"

CHAPTER TWO

THE COMANDANTE AND THE CAPTAIN

The Sea, once it casts its spell, holds one in its net of wonder forever.

—Jacques-Yves Cousteau

The Undersea World of Jacques Cousteau

With a warm smile, President Ronald Reagan welcomed Captain Jacques-Yves Cousteau and his son and fellow ocean explorer, Jean-Michel, into the Oval Office. A few minutes earlier, the president had placed an elegant white-trimmed blue ribbon around Captain Cousteau's neck, supporting a large golden ring upon which was affixed a golden star with white enamel, atop a red enamel pentagon, and with a blue enamel disc in the center bearing 13 gold stars. The captain had just been awarded the Presidential Medal of Freedom on that spring day in 1985. Along with the Congressional Gold Medal, it is the highest civilian award of the United States.

I never had the privilege of meeting Jacques Cousteau, but I have been fortunate to know Jean-Michel, his son Fabien, and his daughter Céline, as well as the children of the late Philippe Cousteau, Philippe Jr. and Alexandria. Like so many of my generation, the name Cousteau evokes joyful, inspired memories of the Cousteau TV documentaries depicting an undersea world of mystery that few had ever seen, let alone visited. The beloved *Undersea World of Jacques Cousteau* series was a major departure from the typical documentary style of the time. Replacing dry, scientific narration was the voice of Cousteau himself, with his elegant, French-accented poetic narrative that profoundly awoke our imaginations and emotions. In her book *Kraken*, author Wendy Williams credits Cousteau with transforming the octopus from a feared undersea monster to a cherished and wondrous creature during one masterfully filmed and poetically narrated sequence. Cousteau went for the heart, not the head, and a generation grew up dreaming of sailing with him aboard *Calypso*, wearing the crew's trademark red watch cap, disappearing over the horizon in search of unexplored seas.

So it's no surprise that I was in tears and barely able to speak when, in front of a small audience in New York, Céline handed me a tall glass sculpture bearing the inscription, "Ocean Advocate Award 2011" as Jean-Michel and Fabien looked on, part of an event Céline had organized honoring her grandfather's 100th birthday. I felt inadequate in expressing my gratitude and reflected on how so many millions around the world have been influenced by the Cousteau legacy. I recalled that during my days at Seacamp, one teenage diver placed yellow tape down the edges of his wet suit, mimicking the original wet suits worn during the Cousteau expeditions. We aspired to be like the Cousteau explorers. So, it has been heartening to see that the Cousteaus continue the family tradition of ocean exploration and conservation today.

Jean-Michel, recalling that day in the White House, sat with me and described their meeting with President Reagan. With the formal ceremony over, the mood was much relaxed as the three made themselves comfortable in the Oval Office. "Dad and President Reagan had many meetings long before that when he was governor of the state of California, and I had the privilege of being there and met with him," he began.

As they sat, Reagan broke the ice. "He said, 'So, Captain. Where are you coming from?' " Jean-Michel smiled, recalling his father's response. "He said, 'Well, I'm coming from Cuba.' It was funny to see the reaction of President Reagan because at the time, of course, Cuba was a major issue. And then at the end of the conversation, after they had recovered and were very social, Reagan said, 'Well, where are you going next?' And my father said, 'Well, I'm going back to Cuba!' and President Reagan was just blown away. It was a very funny situation."

The Letter

Nearly 30 years later, our captain watched with some consternation as an unidentified vessel, gray with no markings, headed straight toward our vessel, which was anchored more than 50 miles off Cuba's southern coast. Others in the crew speculated nervously about the approaching boat, never previously seen in these parts. The boat pulled alongside, and two imposing figures boarded, both in olive military uniforms: a representative of the Ministry of Interior and his taller colleague whose uniform, like the boat that carried them, bore no markings at all. A sidearm hung imposingly from his belt. He turned to the captain and requested to meet with Bobby Kennedy Jr.

At that moment, Bobby—a leading environmental activist, board president of Waterkeeper Alliance, son of the late senator Robert F. "Bobby" Kennedy, and nephew of President John F. Kennedy—was 90 feet below the surface with me and the rest of our group, observing a dozen or so Caribbean reef sharks tracing mesmerizing circles about us. We were carrying the flag of the Explorers Club, documenting previously unexplored coral reef ecosystems in Cuba's southern waters.

After returning to the boat, the mission of our mysterious guests was revealed. We had been visited by a representative of former Cuban president Fidel Castro's personal guard who had a letter from the comandante for Bobby. Mission complete, they posed for a quick photo and departed on the 50-mile journey back to shore and the six-hour drive back to Havana. They had traveled an incredible distance to find us for the sole purpose of hand-delivering a letter to Bobby Kennedy, so we were obviously quite curious as to its contents.

A few days earlier, Bobby and his family had visited with Fidel Castro, who welcomed them warmly. Such a meeting seemed entirely unlikely given that nearly 52 years prior, Bobby's father, serving as U.S. attorney general, and his brother, President John F. Kennedy, were within a whisker of war with Cuba and the Soviet Union during the Cuban Missile Crisis. The quiet Castro–Kennedy Jr. meeting was historical. Relations between Cuba and the United States were warming, though the dramatic announcement of normalization of diplomatic relations would not occur for another six months.

Bobby shared the letter with me, a polite set of Castro's reflections on their meeting and kind words for Bobby and his family. What I found especially significant was his discussion on the oceans:

> For many years I was a passionate spearfisherman without the proper awareness of the beauty and value of coral reefs. Through this I knew some of the experiences of Jacques-Yves Cousteau, who in such a way fell in love with the sea that ended up becoming one of the most famous defenders of the life and the value of the seas. Today it is known that the sea is one of the largest and varied sources of protein foods. These factors helped me understand the importance of the services you have rendered to the people of the United States and other nations of the world in their struggle to protect the environment.

Cousteau's Cuban Legacy

The influence of Cousteau on Castro has been a recurring theme I have heard from Cuban colleagues during my many years working in Cuba. Castro read and was influenced by Cousteau's books, and in 1985 when Cousteau visited the island to make a documentary, the two finally met and shared a special friendship. Fidel granted Cousteau with a rare privilege during his visits. Cousteau and his team became the first non-Cubans to pass through the gate of the U.S. Navy's Guantanamo Bay installation since 1962. Castro is reported to have freed dozens of political prisoners at Cousteau's request. And he spent a great deal of time with Cousteau, dining with him aboard his vessel, *Calypso*.

In the Cousteau documentary that came from that 1985 visit, Cousteau emerges from the water after one dive and climbs the ladder to the awaiting deck of *Calypso*, wearing a grin from ear to ear. He was positively giddy, laughing and shaking his head in disbelief as he bantered with his fellow explorers, themselves excitedly amazed at what they had just seen. Before his visit to Cuba, he had already seen the degradation of reefs in the Caribbean, especially following the loss of *Diadema*, but here he and his team had experienced vibrant, magnificent coral reefs. "My first dive in the waters of Cuba serves as a moment of truth . . . around me, large fish among flourishing coral, a reef more rich than any I have seen in years," Cousteau narrates, a stunning reminder that the unraveling of coral reef ecosystems in the Caribbean was well under way, and even in the mid-eighties it was clear that Cuba was different, spared the demise of ocean ecosystems observed throughout the Caribbean.

Jean-Michel recalls the fact that Fidel was already a skin diver before he met his father and had already protected at least one small part of Cuban waters, a private island on Cuba's southern coast where "Fidel could go diving whenever he wanted to, but nobody else was allowed to go there. I was very impressed by the fact that he wanted to protect it. Of course, maybe for himself. And, I had the privilege of going there. The diversity of marine species in that part of the ocean was very high, very special, and very beautiful." With a note of resignation in his voice, Jean-Michel couldn't help but compare the health of Cuban waters to those of other Caribbean islands he had seen. It was clearly a surprise to him that Cuba's coral reefs were so healthy.

Before allowing the *Calypso* to depart Cuba's waters, Castro challenged Cousteau, asking him why he didn't have a Cuban scientist aboard. Cousteau responded, "Why not?" and later welcomed Dr. Gaspar González Sansón, former vice director of Havana's Centro de Investigaciones Marinas, Center for Marine Research (CIM), to serve as a visiting scientist aboard *Calypso* in New Zealand. Years later, Dr. González would become a dear friend and co–principal investigator for a decade of expeditions off Cuba's northwestern coast. A faded snapshot of Jacques Cousteau and Gaspar standing on the bow of *Calypso* was pinned to the bulletin board in his office.

Gaspar regaled us with hilarious tales of a lonely Cuban among Frenchmen aboard *Calypso*. Gregarious and funny as hell, he delights in storytelling, including his experience aboard the *Calypso*. One night, craving a midnight snack, Gaspar quietly made his way to the galley. He concocted a *Calypso* version of a favorite Cuban treat—cheese topped with guava paste. With the nearest guava paste thousands of miles away, a dollop of jam would do the trick. He sat in the darkness, savoring his creation, when suddenly the galley lights came on and the French crew members stared at Gaspar in horror. "They were furious that I was ruining their fancy French cheese! It was sacrilegious!"

Coincidentally, one of Gaspar's students, Julia Azanza Ricardo, whom I had met when she was a graduate student, told me that she, too, had an encounter with Jacques Cousteau and it changed her life forever. Spirited, funny, and always laughing, Julia recalled her experience as an eight-year-old: "I went to France because my parents were working there for about three years [as diplomats] and we had the opportunity to go the oceanographic museum of Monaco, led by Jacques Cousteau." Her eyes sparkled, recalling her visit that day. "It's a wonderful institution, it's a beautiful museum and I was fascinated about all the creatures. We had the opportunity to meet Jacques Cousteau because we were bringing him a gift from Fidel Castro."

Julia was given the honor of handing the gift to Cousteau. She looked up and recalled how impossibly tall he seemed as she outstretched her arms with the gift box. Though they hadn't met, Cousteau was no stranger to Julia. She had seen his books and learned about his work. "I was fascinated about everything that has to do with his devotion, marine biology."

I asked her what was in the box. "It was a *guanabana*." Julia giggled as we struggled to translate. "Ah, a spiny blowfish! So, it was a dried, inflated spiny blowfish?" I asked. Julia confirmed that's indeed what it was. "Did he like it?" I asked. "Yes, I think so." We both laughed. "I don't remember that. I was thinking about everything else."

Her smile faded slightly and her expression became reflective. Julia looked at me as a wave of emotion welled up within her. "It was like the most amazing experience of my childhood and it helped me a lot to determine my profession. It showed me what I wanted to do when I was a . . . 'grown-up'?" She was unsure of the word. "Yes, 'grown-up,' that's right," I said, and we both began laughing again.

The friendship between Cousteau and Castro continued and strengthened in environmental solidarity at the Rio Earth Summit in 1992 where Castro delivered a sharply worded and uncharacteristically brief address, imploring the developed world to "stop transferring to the Third World lifestyles and consumer habits that ruin the environment. Make human life more rational." In early 1998, less than six months after Cousteau passed away, Castro fondly remembers a playful encounter with Cousteau at the Rio Earth Summit: "They have all the heads of state lined up for a 'photo op' in Rio, and I pulled him [Cousteau] up with me, and say, 'Captain, join this picture in the "photo op" because most people here know nothing about the environment.' And he came up and was in the 'photo op' with all of us." So in that historical photograph of the world's heads of state appears Captain Jacques Cousteau. Though not an official head of state, he appropriately stood for the 70 percent of the planet that had no representation.

If you search Wikipedia for significant events that happened on June 11, you'll find nothing for 1997. The Michael Jordan–led 90–88 victory of the Chicago Bulls over the Utah Jazz that day didn't rise to the level of a Wikipedia entry, but something else should have. That afternoon, something monumental took place on a large island just 90 miles off the Florida coast: Cuba's National Assembly of People's Power passed a remarkable piece of legislation known as Law 81 of the Environment.

The legislation created Cuba's environmental ministry and elevated Cuba to an environmental leader. The law is based on a foundation of exceptionally powerful tenets, including formal recognition "as a basic right of society and its citizens, the right to a sound environment, and to the enjoyment of a healthy and productive life in harmony with nature." The 1992 Rio Summit showed that Fidel Castro had moved decisively green and Law 81 followed.

The law is a truly impressive set of policies and regulations meant to reverse environmental damage from prior decades and chart a path of sustainability. Within a decade, Cuba banned the destructive fishing practice of bottom trawling from its waters. Today, Cuba has nearly met its goal of protecting 25 percent of its marine waters in marine protected areas, one of the largest percentages in the world. (In comparison, the world average is currently 2 to 3 percent.) Many Cubans attribute Law 81 and Cuba's ongoing commitment to the environment to Fidel Castro's environmental ethic, which the comandante, in part, attributed to Cousteau.

In the late nineties, aboard another research vessel visiting from the United States, Castro reflected on his friendship with Cousteau and proudly stated, "You know, he loved exploring Cuban waters because of our protection."

By the time his letter reached Bobby Kennedy's hands, it had been some time since Fidel Castro had last donned a mask and personally explored Cuba's waters, but it was clear that his passion and curiosity for the sea was as strong as ever. In his letter, Castro made a simple but urgent request of Bobby Kennedy Jr: "Today, I beg you, if you have a few minutes, tell me about the general impression of what you have seen on the bottom." Several weeks later, Bobby complied and assured the comandante that for now, Cuba's marine ecosystems were still healthy and spectacular.

CHAPTER THREE

LOOK SMALL

It has long been an axiom of mine that the little things are infinitely the most important.

—Sherlock Holmes

The Scream

Donna Coffin let out a bloodcurdling scream from across the narrow channel of Sawyer Key in the Florida Keys backcountry where a small group of us were snorkeling on our day off. She ripped her mask off, her face white with terror. Startled, the rest of us sprang into action, ready to help. Donna looked up, her expression of raw terror melting into a huge smile and an embarrassed laugh. "It was a barracuda about this big," she laughed as she held her thumb and forefinger about three inches apart. She had spent the past 15 minutes with her face in a small tide pool, less than a foot deep, doing what we taught our students at Seacamp: to "look small," that is, to shift our focus from the large and obvious to the tiny and inconspicuous, for it is there that some of the most fascinating—and important—things happen on our planet. When you change your perspective in this way, a world existing at an entirely different scale starts to take shape. In those few short minutes, Donna was transported to this tiny world where a baby barracuda is transformed into a life-threatening terror.

More than 70 percent of our planet is wrapped in a living saltwater universe, a place that continues to awe and surprise us, a place that is responsible for keeping humanity alive, and yet a place that inexplicably remains mostly unexplored. It is a kingdom of maddening contrasts, from the largest creature that has ever lived—the blue whale—to microorganisms that lie unseen in the dark depths or sweep across vast distances on unseen currents. And so many more that lie undiscovered by humankind.

We humans gravitate toward the big stuff. Travel brochures tout magnificent whale-watching, observing pandas, and swimming with sharks. You won't easily find a travel brochure that summons prospective tourists with the promise of close-up encounters with microscopic dino-flagellates. But consider England's majestic and brilliant White Cliffs of Dover, whose grandeur has drawn humans' awe and reverence for centuries. They have borne witness to some of England's most important history, most notably the return of British soldiers from Dunkirk in World War II, the soldiers' first glimpse of British soil following their harrowing ordeal at the hands of the Nazis. But "look small" at those magnificent chalk cliffs. They're made up of billions upon billions of coccolithophores, single-celled algae festooned with plates of calcium carbonate, which give the cliffs their brilliant white color. Coccolithophores are but one species of phytoplankton, and you can thank your favorite phytoplankton for every other breath you take. They are responsible for half of the oxygen in our atmosphere while simultaneously removing carbon dioxide. They are exquisite, colorful jewels of bewildering shapes and configurations. Some produce hypnotic light in the darkness. Phytoplankton represent the base of the food chain, ensuring survival for even the largest of the oceans' inhabitants. Blue whales feed upon krill, tiny crustaceans, which feed on phytoplankton. Each generation of Peru's legendary sardine population depends upon the phytoplankton they consume.

But perhaps nothing demands one to "look small" as much as coral reefs. They are a layer cake of impossible complexity that science still struggles to comprehend. They are a home to life at every scale, from the microscopic to the colossal. The secret to observing such a diversity of life while scuba diving is simple: Let the divers who look at scuba diving as an endurance sport race over the reef and disappear into the distance while you stay immobile, observing a small slice of reef. Little by little a colorful living community inhabiting the reef emerges before your eyes. Tiny shrimp probe the water with their antennae; small crabs snatch scraps of algae from rocks; tiny plankton float by, eagerly plucked from the current by the tentacles of corals, while tiny fish scamper atop the coral's surface.

Cradled in the Bosom of the Waters

My unexpected love affair with coral reefs began on my 15th birthday when my parents granted my wish for scuba diving lessons, an obsession inspired by reruns of the TV show *Sea Hunt* that ran from 1958 to 1962, starring Lloyd Bridges as Mike Nelson, a scuba diver whose life was filled with adventure and danger. I was attracted by the gear, the adventure, the machismo, and, of course, the huge dive knife he packed. Receiving my certification after a checkout dive in an ice-cold, water-filled quarry near Reading, Pennsylvania, I was eager to continue scuba diving, though perhaps in warmer waters, and I ended up at Seacamp.

Despite the card that said I was a certified scuba diver, Seacamp needed to be sure, so another checkout dive lay ahead on my second day. Our destination was Looe Key, a coral reef six miles due south of Big Pine Key. The waters we plied were glassy calm, like a mirror that stretched to the horizon, broken only by the splash of a small fish. The sky above and the small cumulus clouds were reproduced flawlessly on the water's surface. I peered down at the water and, to my astonishment, could see all the way to the bottom, something alien to a northerner accustomed to the green-brown waters along the New Jersey coast that I observed fishing with my father. We passed over seagrass beds, schools of fish, coral heads, and countless other shapes and colors belonging to objects I couldn't begin to identify. As we approached Looe Key, the bottom was much closer and mountains of mustard-brown corals touched the surface. Our small wake was enough to cause tiny waves to break upon passing them. The corals were arranged in parallel fingers, their tops scratching the water's surface, stretching into the white sands 37 feet below.

Once in the water, we descended slowly down the anchor line, which was visible its entire length, to the anchor that lay in the white sand below us. Soon I was kneeling in the soft, sugar-white sand at the bottom as we gradually formed a circle. Bright sunlight streamed through the surface, illuminating an eruption of color and a tantalizing sense of mystery. I felt a warm embrace of tranquility. I was at peace, joyful and in awe at my extraordinary surroundings, a universe away from the cold green waters of the north I had grown up with. I was wholly unprepared as stunning multicolored fish passed slowly within reach. I marveled at the towering jetties of coral around us, living mountains of corals upon corals, brown and mustard-colored rock-like structures, encrusted with brilliant red, violet, and orange coralline fans and branches, swaying in the warm, nourishing current and, like eager spring blossoms, reaching toward the dancing sunlight scattered on the surface above.

I recognized fish but had little idea of what else I was looking at. I knew the huge jetties or fingers on either side of our group were comprised of corals. But what exactly were they? They seemed to be animal, mineral, and vegetable all at the same time. It was soon apparent that these enormous structures were just the foundation of a rich ecosystem that was exploding with life. An astonishingly diverse collection of fish, invertebrates, algae, and all manner of colorful corals crowded its surface and every crevice for space. Some, like fish, sought protection while others, like sea fans, found a holdfast upon which they could wave back and forth in the gentle current, all the while plucking plankton—a coral's sustenance—from the water sweeping by. And upon them and inside each crevice, there were colorful fish, crustaceans, and worms, leaving no space uninhabited or undefended. As I approached the coralline wall to take a closer look, I became aware of a pinching sensation on my neck, my arms, and oddly, my nipples. Alarmed, I looked about, trying to locate the source, but saw nothing. Then, out of the corner of my eye I spied what can only be described as a pissed-off little fish, dark brown, four inches long, and with all the fury of a 2,500-pound bull. I would later learn that my small but mighty assaulter was a dusky damselfish (*Stegastes adustus*), a highly territorial fish, defending his tiny patch of the great reef. At the time, I was perplexed and annoyed. Later I would regard that aggressive little fish with admiration: Defending a coral reef is a noble cause.

Describing the wonder of a coral reef dispassionately is an impossibility. Today's scientific papers are dry, dense, and devoid of emotion. The journey from the first page to the end is often interesting and illuminating, but at another level, joyless. The magnificence of coral reef ecosystems beckons for the writings of 19th-century explorers and naturalists whose exuberant and exquisite writings were equal parts science and poetry. Especially notable was German explorer Alexandar von Humbolt. Describing a small patch of rainforest near the Orinico River in Venezuela, he is overwhelmed at midday by the constant hum of thousands upon thousands of insects around him. Exasperated, he writes, "There are so many voices proclaiming to us that all nature breathes; and that under a thousand different forms, life is diffused throughout the cracked and dusty soil; as well as in the bosom of the waters, and in the air that circulates around us." Such words could have been written for the coral reef that surrounded us that day.

That dive was an emotional awakening for me as a young teenager. I knew then that the fate of coral reefs would be forever intertwined with mine. But never would I have imagined that I would be working for two decades just 90 miles south of where I knelt. But somehow in my gut I did know that I would eventually run out of adjectives to describe the wonders I would see during those years ahead.

Animal, Mineral, or Vegetable?

When teaching or speaking, I usually ask my audience whether coral is an animal, mineral, or vegetable. Don't worry—it's a trick question.

The most correct answer is that coral is an animal, closely related to sea anemones and jellyfish. Corals are composed of polyps and tentacles with which they grab and consume plankton from the passing current. Coral polyps don't live in isolation—they are colonial. Stony corals, like the ones I beheld at Looe Key, are considered reef-building corals, forming vast colonies, each polyp secreting a calcium carbonate exoskeleton, en masse forming the massive rock structures so important as habitat to a vast array of other species. So, in that way, one could say loosely that corals are both an animal and mineral.

And, yes, corals are "vegetables," too. Many tropical corals harbor symbiotic brown algae known as zooxanthellae, which give corals their distinctive coloration, from mustard to green, red, purple, and orange. It's a good deal for both the coral and the algae. Since most tropical reefs grow in relatively shallow and clear water, the zooxanthellae living within corals, bathed in sunlight, use photosynthesis to produce sugars they share with the corals. In return, the zooxanthellae have a safe home where they can thrive under ideal conditions.

Often known as the "rainforests of the sea," shallow coral reefs like those at Looe Key form some of Earth's most diverse ecosystems. They occupy less than 0.1 percent of the world's ocean area, roughly the size of the state of Texas, yet they provide a home for at least 25 percent of all marine species, harboring a dizzying array of fish, mollusks, worms, crustaceans, echinoderms, sponges, etc., many commercially important, many rare, and many yet undiscovered. Corals take on countless other forms besides the stony, reef-building corals. Sea fans and other "soft corals" adorn the bottoms of shallow tropical waters.

Coral reefs are among the most diverse and valuable ecosystems on earth, directly supporting more species per unit area than any other marine environment, including about 4,000 species of fish, 800 species of hard corals, and hundreds of other species. Scientists estimate that there may be another 1 to 8 million undiscovered species of organisms living in and around reefs. Many drugs are now being developed from coral reef animals and plants as possible cures for arthritis, human bacterial infections, viruses, and other diseases—especially cancer. Corals themselves compete for space using a type of chemical warfare against one another, and biochemical and medical research is focused on learning from this process and finding chemicals that could target cancer cells.

Coral reefs provide economic and environmental services to millions of people around the world, estimated at more than $375 billion each year, a remarkable figure considering that coral reefs cover less than 1 percent of the planet. Coral reefs also contribute to local economies through tourism (scuba diving, snorkeling, fishing, resorts, restaurants, etc.), providing millions of jobs and billions of dollars to the economy.

The commercial value of U.S. fisheries from coral reefs annually has been estimated in excess of $100 million, equaled by the annual value of reef-dependent recreational fisheries. In developing countries, coral reefs are especially important, providing roughly 25 percent of the total fish catch—the food resources for tens of millions of people.

Coral reefs buffer adjacent shorelines from wave action and prevent erosion, property damage, and loss of life. Reefs also protect the highly productive wetlands along coasts, as well as ports and harbors and the economies they support. Globally, half a billion people are estimated to live within 100 kilometers of a coral reef and benefit from its production and protection. A 2008 study by World Resources Institute showed that

28

in Belize the value of coral reefs and mangroves to shoreline protection exceeded the value of fishing and tourism combined. This is especially significant given accelerating sea level rise and increases in the frequency and intensity of hurricanes and other storms. It is estimated that coral reefs absorb an incredible 97 percent of wave energy.

One of my joys at Seacamp was piloting a 30-foot flattop boat loaded with 15 students south to Looe Key—that coral reef that changed my life—where I'd drop the kids into the water and observe their amazement, joy, and wonder. Sadly, most people will never visit a coral reef. And for too many of those that do, they are thought of more as a tropical resort amenity than a critically important part of our biosphere. Scientists still find coral reefs a source of endless mystery and scientific discovery. They serve as the oceans' most critical fonts of life and a safe harbor for millions of the world's most important marine species. To our own species, beyond the unfathomable heartbreak of the loss of such exquisite beauty from the earth's surface, losing corals represents the annual loss of billions of dollars from the global economy and the end of a way of life for billions that depend on coral reef ecosystems.

CHAPTER FOUR

THE CORALS TIME FORGOT

Life finds a way.

—Jeff Goldblum as Dr. Ian Malcolm in *Jurassic Park*

Dreadlocks in the Deep

If you don't pay attention while walking through the back hallways and catacombs of the enormous Smithsonian Museum of Natural History in Washington, DC, you will be swallowed into the maze with little hope of finding your way out. You'll pass the odd room full of collection jars containing slimy creatures, or perhaps see a scientist measuring the leg of a gazelle. As the joke goes, there are people in their eighties still wandering those halls in search of the bathroom they set off to find while in their twenties.

Thankfully, Dr. Nancy Knowlton, Smithsonian's Sant Chair for Marine Science, escorted me through an impossibly complex path of twists and turns to locate her office. Her book, *Citizens of the Sea*, had just been released in 2010, and before we sat down to chat, she took a peek at Amazon to check sales and reviews. So far, so good. The book is based on the Census of Marine Life, a mammoth decade-long project involving 540 expeditions and more than 2,700 scientists from 80 nations, taking on the tall task of recording the diversity, distribution, and abundance of life of 70 percent of the planet. More than 2,600 scientific papers and 30 million data records resulted. The scientists described more than 1,200 new marine species, with thousands more waiting in the wings to be identified.

In her book, Nancy curated some of the most interesting and bizarre of the species encountered, colorfully illustrated with images that are the stuff of science fiction, each one imparting a deserved humility of how few of the ocean's inhabitants we have discovered and described. It's been estimated we're only 5 percent of the way there. Contrary to popular belief, scientists can be funny. Famous for his humorous writings, marine biologist Dr. Milton Love at the University of California, Santa Barbara, famously described scientists having an animated discussion about mucus at the dinner table. So it's not surprising that more than a handful of those 1,200 species bear whimsical names, a favorite being the Bob Marley worm (*Bobmarleya gadensis*), a bright orange polychaete worm found thousands of feet down, among mud volcanoes, in the Gulf of Cadiz in the northeast Atlantic. Sprouting from atop the worm are what appear to be dreadlocks.

What it lacks in a whimsical epithet, my favorite creature in the book makes up for in its jaw-dropping level of bizarreness. Plunge 2,000 to 3,000 feet and you might run into the barreleye fish (*Macropinna microstoma*). It's bizarre enough that this creature has a transparent head, but it gets better. What fools people is that they mistake the fish's nostrils for its eyes. In fact, the eyes, two green tubular structures on stalks, are inside the clear head. In other words, the barreleye sees through its own head! And you thought marine biology was boring.

Bob Ballard, professor of oceanography at the University of Rhode Island, is best known for the discovery of the *Titanic* in October 1987, after 75 years of lying in hiding at the bottom of the Atlantic. But talk to him in private and he'll tell you that his most exciting discovery was years earlier, thousands of feet down in the *Alvin* submersible. They encountered superheated water and hydrogen sulfide emanating in great black plumes from the bottom. But what they saw next turned what we thought we knew about life on earth on its head. It was an entire ecosystem—nine-foot-tall tube worms, mussels, blind crabs, and more—thriving on the energy supplied by these "black smokers," thanks to bacteria that could transform the hydrogen sulfide into the sugars that fueled that ecosystem. For those of us of a certain age, our biology teachers were emphatic that life on earth couldn't exist without sunlight. In an instant, Bob Ballard and the other scientists on that expedition proved them wrong.

The deep ocean is filled with mystery and untold life awaiting the first awestruck gaze of human eyes. Little did we know, until relatively recently, that among that life—even in ice-cold waters thousands of feet deep—were corals. These aren't reef-building corals, but delicate and exquisite branching corals. And they are ancient, the longest-lived animals on the planet, some carbon-dated to be 4,200 years old and still alive, witness to the birth of the Bronze Age, the completion of Stonehenge, the reign of the pharaohs of Egypt, the birth of Christ, and the finale of *Breaking Bad*. And just like their warm water cousins, they are home and refuge to fish and countless other species. Warm or cold, corals are home, refuge, and habitat for a multitude of life-forms they share the oceans with. Thousands of feet beneath the surface, where you find coral, you find fish.

Beneath the Deadliest Catch

It would take 40 minutes to descend through complete darkness, piloting a one-person submersible on the first dive of the expedition to more than 1,000 feet. In the galley of the Greenpeace ship *Esperanza* that morning, my fellow pilots and I drew numbered scraps of paper from a coffee mug held by John Hocevar, Greenpeace's oceans campaign director and leader of the expedition. By luck of the draw, I became the first human being to descend into the two largest underwater canyons in the world: Pribilof Canyon and the larger Zhemchug Canyon, both in Alaska's Bering Sea. Alaska's enormous seascape is full of mountains and canyons that continue the drama of those nestled in its heartland into its surrounding ocean waters. Our starting point, Dutch Harbor, is a point along the huge Aleutian Island arc, a chain of volcanoes, enormous seamounts rising up from thousands of feet below. Zhemchug Canyon, the largest on earth, plunges to 8,530 feet, more than 50 percent deeper and twice as long as the Grand Canyon.

In 2007, I was at the helm of the *DeepWorker* submersible invented by Dr. Phil Nuytten, president of Nuytco Research Ltd., the manufacturer of this tiny technological miracle, capable of a depth of 2,000 feet. The sub is equipped with high-definition video, a manipulator arm for collecting samples, and sonar for navigation and is always in contact with the surface through an acoustic transponder and voice communications. Just like a spacecraft, *DeepWorker* uses carbon dioxide scrubbers to allow the air to be recycled, providing up to 80 hours of life support. A typical dive lasts four to six hours. That may seem like a long time, but the time passes in an instant, and all too soon the dreaded transmission comes from topside: "*DeepWorker* 6, prepare the cabin for recover." Fortunately, for those few for whom the time doesn't pass so quickly, the technology to relieve oneself in the sub is much less complicated than that in a spacecraft (at least for a male pilot): An empty Gatorade bottle awaits under the seat. The sub's propulsion is controlled by two foot pedals, freeing the hands for communications, videography, life support tasks, sonar, and buoyancy control. To Phil's credit, the sub is perfect, though there are often so many tasks happening simultaneously that I might suggest that he consider an extra arm or two for the pilot.

Among the many harried tasks of radio communications, life support checks, and preparing video and other equipment for the upcoming bottom survey, I had a rare moment to reflect on where I was, attempting to comprehend an enormous, complex deep-sea tapestry armed with only with the lights of a tiny sub. At 800 feet, squid enveloped the vehicle, some latching on and appearing to try to take a bite. Others gave a menacing dance in front of the lights, issued a blast of ink, and rocketed back into the darkness. Touchdown was at 1,003 feet in Pribilof Canyon. Per protocol, I radioed life support readings to the surface and was cleared to begin my first transect. My lights revealed numerous cod, perch, small sole, halibut, and skates atop the dark brown, muddy bottom. But I still hadn't found what we had traveled so far to find. The occasional king crab would wander by as I continued along my heading, my sonar set to warn me if I was on a collision course with the canyon wall.

And suddenly, there it was: one of the most beautiful sights I have ever beheld underwater. Out of the darkness a shocking burst of bright pink emerged from the mud, defiant of its dark, colorless surroundings. I had found coral. Beautiful cold-water coral. This was soft coral, branching like a small tree, gently waving in the strong current sweeping across the bottom, plucking plankton it carried. I reversed thrusters and attempted to supplement the video I was taking with still images. The sub was buffeted by the current and I could barely maintain position, even with my foot to the floor. Somewhere a stream of profanity can be heard in an archive recording of that dive as our microphones were live during the entirety of each dive and I did not hold back my frustration trying to get the shot. Thankfully, in the end I was successful, and to this day, I continue to show the image in every presentation I give. This beautiful creature's name is *Swiftia pacifica*, an octocoral found in deep cold water all the way south to Chile and Patagonia. Later, reaching 1,052 feet, I surveyed one of the ridges above the canyon's bottom, finding it covered with delicate white sea whips (*Halipteris willemoesi*), another octocoral, some up to three or four feet long, stretching like bamboo toward the surface. And everywhere we saw coral, we saw fish—and other marine life. Just like corals of the tropics, corals here are critical habitat for other species.

"We went on a listening tour across coastal Alaska," explained John Hocevar. "We had been contacted by indigenous communities in the [Bering Sea] islands that had asked us for help. This was unusual for us and uncomfortable for them. They blame Greenpeace and environmental organizations for a lot, like limiting their fishing." John directs Greenpeace's oceans campaign. Greenpeace bore the blame for the end of commercial whaling and sealing, though John explained that the organization never targeted the indigenous populations. But now the communities were facing despair, gravely concerned about giant factory trawlers essentially dragging their nets right up to their beaches. "It had become very difficult for people on these islands to eke out a living, even though the surrounding waters are some of the richest in the world," John continued. "There was a real concern about the vast quantity of fish the trawlers were taking from their waters, while at the same time they were hearing stories and seeing evidence of ancient corals and sponges being dragged up at the same time." What the local people may not have fully understood at first was that the factory ships weren't just taking fish. They were also tearing away swaths of the very ecosystem that allows fish to survive. With more and more corals and sponges landing on the decks of ships thousands of feet above, fish were losing their habitat, facing an increasingly barren seascape.

Thanks to a generous donor, Greenpeace set forth on a scientific expedition with participation by the National Oceanic and Atmospheric Administration (NOAA) to document deepwater corals in the Bering Sea—and document the damage caused by factory ship trawling—in order to secure protection for the Bering Sea Canyons. I was brought on as a project consultant and an submersible pilot.

Several scientific papers were published, including the discovery of a new species of sponge and previously unknown ranges for various species of corals. But most important, John says, "We demonstrated that corals and sponges provide habitat for fish and other types of marine life." He sighs in frustration, reflecting on the fact that industry and fisheries managers refuse to connect the two. "They tend to treat fisheries on a single species basis, and it's one of the reasons why we have collapses of populations without seeing them coming. We can destroy enough of the habitat for a particular fish and it may not even have been that we overfished it necessarily—we just eliminated enough of their habitat so that the population collapsed." Greenpeace continues its efforts to protect the Bering Sea Canyons to this day, with some modest victories but still facing a protracted fight.

It was only a few days after I had seen the most beautiful thing I had ever seen underwater that I saw perhaps the worst. It was the penultimate dive on Pribilof Canyon. The bottom was a bit more than 1,000 feet and I landed on what appeared to be some sort of geologic stratification—unusual layers and grooves of sediment in parallel lines across my path. I couldn't make heads or tails of what I was seeing. I pivoted the sub to illuminate more of this linear structure and suddenly realized what I was looking at. A trawl "scar," deep ridges made by the weights of a trawl net dragged across the bottom of the sea. A wide swath of bottom appeared as if it had been plowed like a cornfield, overturned sediment neatly piled along the long groove. With the corals gone, a lone fish tried to find refuge in one of the grooves made by a wheel of the trawl. My colleague and fellow pilot, Michelle Ridgway, who worked for years in these waters and was piloting a second submersible on this dive, explained to me that some of the trawls used are as wide as a Boeing 737's wingspan. I flew along the scar—it went on much farther than I could follow it—probably miles.

We began our transect, but shortly thereafter I was told to hold position—apparently the squid had won the last round against Michelle, causing one of her thrusters to blow a fuse. She surfaced for an early recovery while I continued the dive alone. I was excited to see a number of corals. The bottom was covered with tiny (an inch or two long) white sea whips, one of the corals we had seen elsewhere in Pribilof Canyon. But the sea whips we had seen previously were much larger, three or four feet long. I only spied two or three that big in this location. I then noticed a strange white ridge along the black horizon. As I approached, this ridge lay directly in my path, straight as an arrow. Sure enough, this was another trawl scar, larger than the first. I radioed to Sasha at the navigation station on the ship and asked that he note this location on the position indicator. As I continued the dive, I found many more linear features along my path, more trawling marks, no doubt, perhaps much older ones, distinguished only by tiny white sea whips, pushing up and trying to make a go of it. They were all roughly the same size, suggesting they were roughly the same age, most likely regrowth after the area was trawled, like even-aged stands of trees in the Pacific Northwest, planted after the old-growth forest has been decimated.

Several years after that expedition, I had a chance encounter with Captain Keith Colburn, captain of the *F/V Wizard*, a crab-fishing vessel featured on the TV series *Deadliest Catch*, at a fancy Washington, DC, dinner for Capitol Hill Oceans Week. As I flipped through the images of corals on my iPhone, Captain Colburn looked on with a combination of interest and skepticism. "Keith, this is habitat. You need to be concerned about it. This is important to fish. This is important to crab." We exchanged a few emails for a while. . . . I still hope to build allies with fishers like him. He remembered when the Greenpeace ship *Esperanza* was in Dutch Harbor, but perhaps the association with Greenpeace is a bridge too far.

At the departure gate at Tom Madsen Airport in Dutch Harbor, I felt rather out of place at the beginning of my long journey back to Washington, DC. I was surrounded by fishers. I struck up a conversation and learned that they had just disembarked from one of the factory ships. Unable to resist a captive audience, I balanced my laptop in my palm and gave an impromptu PowerPoint presentation, scrolling through the images of the exquisite corals we had just documented. "Yeah, we see that stuff on the deck all the time," said one fisher, the others nodding in agreement. I replied, "That's why we're here. We want to help keep this off your deck." I'm not sure they understood the main point. That in the blink of an eye, corals that may have taken centuries—or millennia—to grow are lost and that—whether in the Bering Sea or Caribbean Sea—the demise of corals is the demise of the ocean's bounty . . . and the oceans as we have known them.

CHAPTER FIVE

THE PRINCESS AND THE PEA

How did it get so late so soon?

—Dr. Seuss

Mission to Mars

I waited in the tiny airlock as my hosts crowded in. One of the team cranked the wheel on the hatch and with a metallic report it was sealed behind us. Turning the wheel on the inner hatch, he slowly released the seal and warm, humid air rushed in and the musty, pungent aroma was an unmistakable signal to my senses that we had been instantly beamed to a tropical rainforest. Except my senses were wrong. It was 2002, and we were standing in the middle of the Arizona desert, inside Biosphere 2, a massive sealed ecosystem under glass covering more than three acres. It was built by a private venture as an experiment that sought to replicate many of Earth's natural systems to support human life in space for long space journeys, such as traveling to or colonizing Mars.

In the original experiment, seven humans lived inside for two years, living solely off the food they grew, the oxygen produced by plants, and recirculated water. I worked with some of those original "Biospherians" when I codirected a study of carbon cycling in Biosphere 2. The Biosphere 2 experiment was run as if it were a space mission, complete with a nearby mission control center that monitored and controlled Biosphere 2 as it "traveled" 1,000 miles per hour, completing one rotation around the planet per solar day while orbiting around the sun once per solar year, the identical velocity and trajectory as the Eiffel Tower, Empire State Building, or nearby cactus. Biosphere 2 was planted on the desert floor, but it was fun to imagine it as a full-fledged spacecraft hurdling through space.

Never before had I set foot inside Biosphere 2, which included a million-gallon ocean fringed by a small sand beach, a rainforest, a savanna, and an agricultural zone to grow crops and raise a modest quantity of poultry and livestock. Its size was overwhelming, a marvel of architecture. Soon after Biosphere 2 was sealed with the original crew of seven inside, oxygen levels plummeted to the equivalent of 17,000 feet above sea level. Carbon dioxide levels skyrocketed to near-toxic levels (the latter being the reason we were brought in—to understand why). Weeds overran food crops. Only 6 of 25 species of vertebrates survived. Most flowering plants died since bees and other pollinators also died. In the end, cockroaches and ants dominated, and they weren't even supposed to be there. For the human inhabitants, life was a hungry, brutal struggle. (We were forbidden to mention food in our communications with the Biospherians.) Through the glass, I spoke to Biospherian Abigail Alling with a telephone handset. She was ghostly pale and shockingly thin, yet she still managed a smile during our brief visits. Biosphere 2 was a stunning lesson that even in a controlled, state-of-the-art environment, we don't know how to engineer a system that duplicates our natural ecosystems. As ecologist David Tilman stated, "That should serve as a warning to those who think we can restore ecosystems on Earth once they have been dismembered."

The lesson of Biosphere 2 is one that is especially resonant to those who care about the oceans. We are in the midst of a large, uncontrolled experiment on our oceans, facing great uncertainty as to whether today's impacted ocean ecosystems can be restored as we continue to tear at them. The difference—a profoundly important one—is that we're tinkering with Biosphere 1. And when it comes to coral reef ecosystems and the dizzying complexity of relationships among its thousands of inhabitants and other species that rely on them, many still unknown or little understood, the lesson is especially compelling and humbling.

One Fussy Invertebrate

Not only is the list of factors that make coral reef ecosystems a marvel of diversity and complexity a long one, so, too, is the list of factors that can either keep them healthy or make them sick. Like Hans Christian Andersen's "Princess and the Pea," corals can be extremely sensitive and seemingly impossible to please. Just about every human-caused activity, it sometimes seems, can represent a major threat to coral reefs, attacking them at both a local and a global level.

Some threats are pretty obvious, like fishing with dynamite or poison, still a practice in parts of the Indo-Pacific. Some are less obvious, like spreading Miracle-Gro on your lawn on a Saturday morning. It's a long list for a fussy invertebrate. The list goes on and on: coastal development, erosion, dredging, deforestation, sewage discharge, cows defecating in rivers, disease, tourists, acid oceans, African dust, sunscreen, anchors, ship groundings, plastic bags and other trash, and killing coral for jewelry or souvenirs. There are a troubling number of peas under that mattress.

Of course, the most iconic impact making headlines more and more is coral bleaching, almost always the result of ocean waters that are increasingly warming to the point that corals are weakened or die. Bleaching events have increased in frequency and are covering larger and larger areas. And despite the urgent call for meaningful action that has echoed for decades, the world continues to fall short of its targets, and the impacts to Biosphere 1, both seen and unseen, continue to mount.

Disaster in Slo-Mo

As a filmmaker, Jeff Orlowski has a keen skill of being in the right places at the right times. Or perhaps better put, the most horrific places at the right times. He directed the 2012 Emmy-winning documentary *Chasing Ice*, following National Geographic photographer James Balog in his Extreme Ice Survey across the Arctic, deploying time-lapse cameras to capture a multiyear record of the world's melting glaciers. The resulting videos, compressing years of still images into seconds-long videos, reveals the haunting and unmistakable impacts of a warming climate as massive glaciers literally disappear before one's eyes. In one real-time sequence, the team caught the unimaginably massive collapse of a glacier, the scale of which challenges human perception to comprehend. Similarly, our brains are poorly wired to process changes occurring over decades or centuries. Such changes can be nearly imperceptible, if noticed at all. Our cave-dwelling ancestors, from whom we inherited our brains, pondered life day by day. Wake up, forage for food, avoid being killed, and on a really good day, reproduce. Neuro-economist George Lowenstein puts it this way, "Our emotions are like software programs that evolved to solve important and recurring problems in our distant past. They're not always well-suited to the problems we face in modern life. It's important to know how our emotions lead us astray so we can find ways to compensate for these flaws." In his book *How We Decide*, Jonah Lember elaborates on these emotions: "Because the emotional parts of the brain reliably undervalue the future—life is short and we want pleasure now—we all end up spending too much money today and delay saving until tomorrow . . . and tomorrow, and tomorrow." So it makes sense that we have trouble wrapping our human heads around slow-motion processes like global warming, which has unfolded for over more than a century and will continue to do so for decades to come. In dramatic fashion, the film's time-lapse sequences proved an incredibly effective tool in helping viewers overcome the brain's limitations, able to see with their own eyes the dreadful impacts of a rapidly warming climate unfolding across three continents. It has changed more than a few minds among so-called climate deniers.

With the success of *Chasing Ice*, Jeff and his team pondered a sequel to document another striking example of the impacts of climate change, this one down under, in Australia, on the Great Barrier Reef. Using the same time-lapse techniques used to document the melting of glaciers, *Chasing Coral* set out to document the widespread and devastating impacts of coral bleaching, the direct result of warming oceans, which is a direct result of climate change. Many corals are already near the top of their thermal tolerance, the hottest water that they can stand. Once corals become too stressed by temperature, their symbiotic relationship with zooxanthellae—those algae that give corals their coloration and nourish them with sugars they produce via photosynthesis—breaks. The corals expel the zooxanthellae, which leaves them ghostly white and in a seriously weakened condition, though still alive. Corals can recover from bleaching, but prolonged warming events—which have become increasingly frequent—push corals over the edge. More than ever before, coral bleaching is becoming coral death. Jeff recalls the impetus behind documenting coral bleaching. "There was a sense that people didn't realize how bad it was, and in many cases, people didn't know about coral bleaching in the first place. Our intention was to visualize climate change and the hope was to leverage the corals to visualize climate change."

In 2015 there was a buzz among climate experts, oceanographers, and other scientists. Something big was coming. The unavoidable conclusion was dire: An enormous ocean warming event was on its way and, among other places, would strike Australia's beloved Great Barrier Reef early the following year. Again, Jeff's uncanny timing set the stage. The film crew and scientists mobilized and began deploying underwater versions of the time-lapse cameras used in the prior film. Little could they imagine the devastation to come. They would be documenting record-setting warm sea surface temperatures and the most widespread and severe bleaching event ever recorded on the Great Barrier Reef. At the time, scientists estimated that nearly one-third of the corals had been killed by the event. Some estimates ran higher. The time-lapse sequences captured by the team were at the same time mesmerizing and nauseating as the exquisite and brilliant rainbow of colors in the vast community of corals was suddenly extinguished. The stark white skeletons of the corals remained—for a time. Many were quickly overgrown with algae. Many began to break apart and erode from existence. The contrast couldn't have been more profound.

For Jeff and his team, it was paradoxical. "We were able to pull it off and we stayed through, and fortunately—" Jeff interrupted himself. "It's so odd as well because I just said 'fortunately.' You experience really weird emotions around trying to capture a terrible thing. I think that it's been a really complex set of emotions for myself and our whole team. It's emotional burnout on the state of the planet. That's been real." As the film was nearing its release in 2017, another mass bleaching event would strike the Great Barrier Reef. Such back-to-back events were unprecedented, as was the toll. A study published in late 2020 based on work by marine scientists at the ARC Centre of Excellence for Coral Reef Studies in Queensland found that Australia's Great Barrier Reef had lost more than half of its corals since 1995 due to warmer seas driven by climate change. Long upheld as the gold standard of healthy and well-managed coral reefs, such devastation to the Great Barrier Reef seems to many unimaginable and stands as one of too many lessons reminding us of the vulnerability of coral reef ecosystems and the devastating power of a warming climate to destroy them.

Unfortunately, climate change isn't content with inflicting bleaching on corals. It also seeks to tear them apart, limb by limb. Higher sea surface temperatures are believed to be responsible for a substantial increase in Atlantic hurricane activity since the early 1980s, including intensity, frequency, and duration, as well as the number of strongest (category 4 and 5) storms. Hurricanes can inflict substantial physical damage on coral reefs, especially branching corals like elkhorn coral. Coral reefs and hurricanes have coexisted for millennia, and corals can recover from a storm event. But the damage inflicted by stronger storms can be significant, and more frequent storms means corals may have less time to recover from one storm before being hit by another. And while in the past corals could recover from hurricanes, the combined effect of other human impacts can mean that a hurricane today is striking a coral reef already weakened by other factors.

Jeff expressed his gratification in the fact that *Chasing Coral* had a strong impact and noted a change since the film's release. "I do feel like the zeitgeist is a bit different now in 2021, compared to 2017, where, just four years ago, we still needed evidence of climate change. And now only four years later the notion of making a film that is evidence of climate change seems moot. The conversation has completely shifted to how fast we're moving on solutions. This is real, this is happening, and what are we going to do about it?"

Nancy Knowlton spoke by Zoom to my class from a laptop perched on the dashboard of her Tesla parked at a recharging station along I-95. "It's really been clear since about 1998, with big El Niño bleaching events, and then increasingly since then, that we really can't expect to have healthy coral reefs in the absence of tackling climate change. So it's really important to take these local actions because the reality is we lost most of our coral reefs, at least in the Caribbean, before global warming was an issue."

There is no doubt that Cuba is in hot water along with the rest of the world. On a dive in June 2016, I happened to glance at my temperature gauge. At 30 feet, the temperature in these Cuban waters was nearly 90°F. But while Cuba has experienced coral bleaching, even in its healthiest protected areas, the bleaching recovers rapidly. The few bleached corals I observed that June—mostly in the genus *Agaricia*, a platelike coral residing in deeper waters—had fully recovered by November. My Cuban colleagues explained that this had been a typical pattern for years. I couldn't help but think that these corals were bathed in the same hot water as those in the Florida Keys, just a short distance away, yet the Keys had shown repeated widespread bleaching events. Research released in 2021, focused on Cuba's Jardines de la Reina, Gardens of the Queen, would begin to offer reasons why, including the fact that these corals are healthier to begin with due to local practices, and are presumably more resilient. Meanwhile, for an American it is a strange world indeed to be in a country where there is no political divide on climate change, indeed no divide on environmental issues at all. Cuban scientists are respected, not ridiculed, as has been the unfortunate recent trend in the United States. And so what would be unthinkable in the States came to be in 2019 when Cuba updated its constitution and, with widespread public support, became one of only 10 nations worldwide to enshrine the fight against climate change in its constitution:

> Article 16 (f) promotes the protection and conservation of the environment and the confrontation of climate change, which threatens the survival of the human species, on the basis of the recognition of common responsibilities, but differentiated; the establishment of a fair and equitable international economic order and the eradication of irrational patterns of production and consumption.

Admittedly, reducing the island's greenhouse gas emissions would not amount to much on a global scale; for example, in 2014 Cuba was responsible for 0.1 percent of the world's total carbon dioxide emissions (versus 15 percent for the U.S.). Yet it is significant for Cuba to lower its reliance on petroleum for electricity generation. The Caribbean is the region of the world with the least efficient and most expensive means of electricity generation: diesel-fueled power plants. Cuba is also experiencing sea level rise, evidenced along the north coast of Cuba not far from Havana where beach dwellers are spreading their blankets upon the ruins of homes that have surrendered to the waves. Cuba is now forbidding construction of homes in the coastal zone and actually razing homes and relocating residents inland. Meanwhile, in the United States, after homes are destroyed by hurricanes and floods, owners often build back in precisely the same spot, thanks to federal aid and insurance companies.

The Evil Twin

We already know that an increasing concentration of carbon dioxide in the atmosphere leads to global warming, including warming oceans. Unfortunately, carbon dioxide has another trick up its sleeve that's not only terrifying for the fate of corals but for any of the ocean's creatures with a calcium carbonate shell. Combine carbon dioxide (CO_2) with water (H_2O) and the unfortunate result is carbonic acid (H_2CO_3). Acids of any kind gladly eat away at calcium carbonate—such as the skeleton of reef-building corals—and therein lies what some refer to as climate change's "evil twin," though perhaps more accurately, "eviler twin": ocean acidification. While its potential impacts are profound, the fact that fossil fuels are acidifying our oceans is much more rarely discussed, and the existence of this grave threat is not well known among the public. Yet many put the potential harm to ocean life by acidifying oceans at the same level as warming oceans. Coral reefs face the indignant end of simply dissolving away.

Since the beginning of the industrial revolution, the pH of surface ocean waters has fallen by approximately 0.1 pH units, representing approximately a 30 percent increase in acidity. Future predictions indicate that the oceans will continue to absorb carbon dioxide, further increasing ocean acidity. Estimates of future carbon dioxide levels, based on business-as-usual emission scenarios, without reductions in carbon dioxide, indicate that by the end of this century the surface waters of the ocean could have acidity levels nearly 150 percent higher. Not since the Miocene epoch more than 20 million years ago have the oceans seen such a level of acidity.

Acid oceans not only threaten corals, but also clams, oysters, mussels, conchs—anything with a shell. But most chilling is the fact that most of the sea creatures with shells are at the base of the marine food chain. Those billions of coccolithophores, single-celled algae covered with plates of calcium carbonate that make up the great Cliffs of Dover, lie at the base of the food chain. And there are many others keeping them company— both algae and animal.

One might think that an unlikely authority on the subject would be a U.S. senator. But Senator Sheldon Whitehouse, a Democrat representing the Ocean State, Rhode Island, is no ordinary senator. He's married to a marine biologist, Dr. Sandra Whitehouse. The senator recalls their unique courtship, which, rather than going to a play or antiquing, consisted largely of freezing his butt off while they dove in the chilly, inhospitable waters of New England. Despite the pain, a man from the Ocean State became even more of an ocean lover—and authority. In 2013, Senator Whitehouse founded the Senate Oceans Caucus, which he cochairs with Republican senator Lisa Murkowsi of Alaska. Whitehouse points out that the cochairs represent the largest state and the smallest state in the United States as well as two different political parties. It's an active and productive caucus that embraces something rapidly evaporating in the United States: bipartisanship. Our paths crossed because of the unusual fact that our respective work is focused not only on the oceans but also on Cuba. Whitehouse has been a leader on improving U.S. policies related to Cuba. Whitehouse and his wife have accompanied me to Cuba to observe firsthand the state of the reefs and learn about collaboration with Cuban scientists.

And so a U.S. senator can more than hold his own should the issue of ocean acidification pop up. "As terrestrial creatures, we're too often ignorant of the damage that's being done in our oceans because very few of us have the ability to see it. And I think there is an important responsibility for those of us who do see it to warn how bad it is and how dangerous it is if the slope really starts to slip. How many people even know what a pteropod is?" he asks. Not many, I would guess. A pteropod is a tiny, free-swimming snail or slug, less than one centimeter in length. "A marine survey off the Pacific Northwest found that 50 percent of the pteropods out there were exhibiting severe shell damage. And severe shell damage is obviously associated with ocean acidification," Whitehouse continues, reciting the chemistry underlying the threat. "The pteropod is at the base of the ocean food chain and you know what happens if we lose the humble pteropod. They're creatures of the Lord's that do immensely important work for our planet and for our species. And we ignore and mock their condition at our peril."

I told Senator Whitehouse that I always thought that ocean acidification would be the issue to foster unity in Congress to take decisive action on carbon dioxide emissions. After all, it's a way to remove climate from the equation and even rally the climate deniers. So far, not happening.

Gift from the Sahara

Taking my first breath of fresh air after deplaning onto the tarmac at Cyril E. King Airport in St. Thomas in 2004, I was bewildered. Having lived for a decade in Southern California, I thought I was seeing smog. Nick Drayton, then the director of Ocean Conservancy's U.S. Virgin Islands office, straightened me out. I was seeing—and my lungs were feeling—part of the Saharan Desert, swept airborne by strong winds, especially during periods of drought on the African continent. The dust carries a fungus, *Aspergillus sydowii*, which attacks coral reefs, while the iron and silicates that comprise the dust fertilize the otherwise nutrient-poor waters, promoting the growth of algae that can smother and slime the reefs. The cloud of dust I saw was far larger than a hurricane, spreading across most of the West Indies as it passed westward. It was a well-known phenomenon for Caribbean islanders, but new to me, and now, in my face—literally. Yet another threat to coral reefs and yet another phenomenon made worse by global warming, causing such dust "storms" to become more frequent and severe.

A Pox on Your House

Over the past decades, diseases attacking corals have emerged as a major threat. Not only are diseases more widespread geographically, but they've become increasingly lethal—presumably due to more vulnerable corals already under assault by other factors. In addition, never-before-seen diseases are emerging with lethal consequences. Especially chilling is the fact that there is now no doubt that infections by human pathogens have contributed to major losses in elkhorn coral; the Florida Keys are a vivid example, suffering from leaking waste from septic tanks built in porous limestone and the reason for a major sewer system being completed for the Keys as part of the massive Everglades restoration project. In particular, the pathogen *Serratia marcescens*, when it infects humans, is opportunistic, causing respiratory, wound, and urinary tract infections. In coral, it causes a disease dubbed "white pox," named for the white blotches or scars that appear on infected elkhorn coral where the living tissue has disappeared, revealing only its calcium carbonate skeleton.

Now, a new, especially lethal disease has spread rapidly throughout the Caribbean since first reported in Florida in 2014. It's called stony coral tissue loss disease (SCTLD). Its cause is unknown, but it is attacking more than 30 species of hard (stony) corals including brain, pillar, star, and starlet corals. Mortality from this rapidly spreading disease is frighteningly high. Infected coral colonies display multiple lesions and die quickly. Outbreaks of SCTLD are documented in Jamaica, Mexico, Saint Maarten, the U.S. Virgin Islands, the Dominican Republic, Turks and Caicos Islands, Belize, the Bahamas, Puerto Rico, the British Virgin Islands, the Cayman Islands, Guadeloupe, St. Lucia, Honduras, Martinique, and elsewhere. The latest map from the Atlantic and Gulf Rapid Reef Assessment (AGRRA), a network of scientists monitoring corals throughout the region, shows a ring of infections encircling virtually the entire Caribbean Basin. But in the middle, Cuba remains untouched. "We really don't have many diseases here," says Patricia González, "not only the number of diseases but also the frequency of diseases." As with climate change, protecting corals locally may well be increasing their resilience and resistance. But, as is true with so many mysteries beneath Cuba's waters, more study is needed.

The U.S. Invasion of Cuba

In August 1992, Hurricane Andrew became one of only four category 5 hurricanes to make landfall in the United States. It roared across Homestead, Florida, 25 miles south of Miami, chewing homes and vegetation along its deadly path, changing the South Florida landscape—and building codes—for decades. The hurricane capsized dozens of vessels, and aboard one of them were visitors from the Indo-Pacific in an aquarium: the beautiful lionfish, *Pterois volitans*, bearing striking bright red and white–striped "candy cane" coloration and an array of long, sharp, threatening spines. For good measure, the spines are poisonous. Whether a mushroom or a fish, such attention-grabbing bright and often red coloration, known as "warning coloration," is an aesthetically pleasing admonishment to stay the hell away. A lionfish sting won't typically kill a human, but can inflict severe pain and inflammation, depending on where victims are stung. Perhaps as few as three of these fish spilled into the coastal waters and went forth and multiplied at an astonishing rate, spreading north and south along Florida's east coast before venturing out into the Gulf and Caribbean.

Well, maybe that's what happened. No one really knows for sure. The hurricane might have taken out someone's home aquarium. Or perhaps Andrew deserves none of the blame and the owner of an aquarium had to move in haste and decided to release their lionfish humanely into a canal rather than a Dumpster. There are other theories, but most agree that ground zero was the Miami area and the timing in the early nineties, though there may have been sightings as far back as the mid-eighties.

With no natural predators in this strange new world, the lionfish gladly settled into the coastal coral reefs and engorged themselves with small reef fish, many juveniles. Some have described lionfish as bulimic, feeding a voracious appetite until their guts are so full—stretched up to 30 times their normal state—that they expel the partially digested contents and begin anew. Stomach contents analyses of lionfish have revealed that they will eat just about any small fish—and the occasional crustacean—that comes their way, which includes up to 50 different species, such as juvenile parrotfish, grouper, damselfish, and shrimp.

They've grown fat and happy—I've encountered giants, nearly a foot and a half in length, hanging upside down in underwater caves. Divers must now learn quickly not only to be conscious of the ceiling of a cave to prevent a bump on the head, but also to prevent a scalp full of poisonous spines. Be sure your underwater light has fresh batteries before poking your head into dark and cozy underwater places!

Here again, like the pathogen that caused the Great *Diadema* Die-Off, an invader from another ocean earned the description "invasive exotic" species, as the invasion was occurring in astonishing numbers and rapidity. Every 55 days lionfish are capable of laying hundreds of gelatinous eggs, and the larvae that hatch drift great distances before settling onto coral reefs where they mature into adults. By the early 2000s, lionfish had spread up the east coast of the Florida peninsula and found their way along Gulf Stream waters all the way to Bermuda. Early hopes to contain the spread were quickly dashed. A few short years later, lionfish stretched as far north as Rhode Island and had enveloped the Bahamas. From there they spread west along the Gulf Coast to Texas, turning south along

Mexico and the Central American isthmus, simultaneously spreading eastward to the windward islands and turning south to invade the South American continent, Venezuela, and Columbia. By 2008 Cuba reported its first sightings along its northern coast. Within five years, lionfish had enveloped the entire massive Cuban island. At present, there are precious few refuges in the Caribbean Basin free of this great invasion.

The presence of lionfish, and their consumption of the juvenile fish that would normally become an integral part of the complex ecosystem that keeps coral reefs healthy, could have long-term effects on the health of coral reefs. Removing so many small colorful fish from the reef, they are also impacting tourism. And they're going deep; they've been found several hundred feet below the surface. I've encountered them more than 100 feet down, fat and happy.

Lionfish are delicious—just remember to remove the spines. I sampled my first at an Explorers Club dinner in New York—white, flaky flesh with a delicate flavor—a chef's delight. Dive clubs regularly hold lionfish "derbies" to spear as many lionfish as possible. Fish markets are encouraged to sell the fish. All manner of fish traps have been tried, but in the end the best control is found in areas easily reached by divers, or in well-managed protected areas where clearing the reefs of lionfish is now a standard part of their duties. Divemasters, working for tourism companies, perform the same function, and it's even become an attraction for visitors, able to spear the only fish permitted to be taken in protected areas. Had we been keeping score, Bobby Kennedy and his son caught so many, they would have won gold medals on one of the trips I led to Cuba.

In the end, though, no one believes that there are enough divers covering enough territory to make a dent in these naturalized citizens of the western Atlantic and Caribbean. In the absence of an Elon Musk–esque technical silver bullet, many of us have concluded that Mother Nature herself must assume the task of eradicating these invaders. The problem is that in most parts of the Caribbean, we have already eaten the large predators—especially sharks and groupers—that might be part of that solution. Worldwide, we have already removed an unimaginable 90 percent of large predators like sharks from our oceans and planted them on our dinner plates. But even in protected areas like Cuba's Gardens of the Queen, the largest no-take (no fishing) protected area in the Caribbean, where these predators abound, none have encountered this exotic species before, so they stay away. Is it possible to convince them that these strange fish are quite edible and tasty? A researcher in Honduras tried to "train" sharks to eat lionfish, with limited success.

In Cuba, with a massive marine protected area (MPA) where sharks and groupers still thrive, it would seem like the ideal place for biological control to take hold. In the Gardens of the Queen, I set off in the morning with a small group of visitors I had brought to Cuba in 2017 to accompany our divemaster, Tony, on a lionfish control dive. He would demonstrate their work interacting with these predators, trying to "train" them to eat lionfish. Within seconds of spearing his first lionfish, a large Caribbean reef shark appeared—and swam right for the tip of Tony's spear where the impaled, wriggling lionfish awaited. The shark enthusiastically pulled the lionfish from the spear and it disappeared down his throat within seconds—perhaps a bit spicy with the spines still attached. Tony, who seems to know every shark on the reef personally, has noted no ill effects on the animals since he started this practice.

Tony speared a second lionfish, and again, within seconds, another shark appeared. Initially uncertain of where his tasty prize awaited, the seven-foot shark swam behind and beneath two of our divers then quickly turned upward directly in front of their masks. A startled young woman raised her hands as if to show the palms of her hands to the shark and say, "Not me. I don't have it!" This vividly demonstrated one of the weaknesses of divers spearing lionfish. The sharks hear the sound of the spear—their dinner bell—and can represent a threat to divers, especially those inexperienced in the presence of these imposing predators. We moved on and spotted a large black grouper hovering over the bottom. The grouper was a bit more cautious and did not approach, so Tony pried a speared lionfish from his spear, allowing it to sink to the bottom within a few feet of the grouper. Still cautious, the fish swam in a circle, assessing this odd-looking wounded fish. Slowly, he extended his enormous mouth and the fish was sucked inside. No sooner had it swallowed the fish, it spit it right out again. I wondered if the grouper had been painfully stung or if it didn't care for the flavor. In fact, the grouper wisely decided that its original plan of eating the lionfish tail-first wasn't the best option and swung around where it vacuumed the fish headfirst down the hatch, presumably to allow the spines to fold in and the fish to glide down its gullet. Our final stop was a hole on the sea bottom out of which a small green moray eel curiously scanned another lionfish that Tony had dropped. Within a few bites and mild thrashing of its head, the lionfish was deep in the eel's gut.

Topside, I asked Tony if these predators would eat lionfish without diver intervention. I was gratified to hear that they had observed a few instances of this happening. But there remained many questions: Were the only predators freely taking lionfish those that had already been "hand"-fed? Would their fellow sharks, groupers, and morays follow suit? It raised interesting questions of animal behavior and evolutionary biology. Of our ancestors, what were the circumstances around the first humanoid to pick that red berry from a bush and eat it? And if they weren't killed or sickened, how long would it take others to learn to eat what we would eventually know as a raspberry? But one thing was certain: If we were indeed to find a way to control lionfish with nature, a healthy predator population would undoubtedly be part of the solution, and we'd need lots more healthy predators. The implications for fishing restrictions and more protected areas are clear, and Cuba may be in one of the best positions to show us why.

The 50-Year-Old Rolodex

In April 2010, I bent over to pick up tar balls along Louisiana's Gulf shoreline more than 60 miles from the Deepwater Horizon platform, a stark reminder of how massive the oil spill was. I was there, in part, to collect water samples to better understand the extent and impact of the spill, but like many nongovernment organizations (NGOs) and academic institutions, we found it nearly impossible to get out on the water. BP had essentially chartered every boat in the area. Going boat to boat I'd ask, "Captain, would it be possible to charter your vessel to collect some samples?" One captain with a BP contract responded, "I'm sorry, Doc. The pay's too good and the work's too easy." For him the "work" consisted of being paid to sit in port, enjoying a beer and chatting with his fellow captains. It's hard to compete with that. Thankfully, Jean-Michel Cousteau and his nonprofit group, Ocean Futures, had dispatched a team to the area. In addition to their Zodiac, they had managed to find a strong-willed boat captain, resolved not to allow himself to be tempted by the dark side—BP—and help us out, albeit for much less pay and for actual work. With my brother Alan video-documenting our efforts, we managed to get the samples and airship them to the lab packed in dry ice.

We witnessed helicopters dropping sandbags along the shoreline for protection, but doing so with such poor precision that there were massive gaps easily allowing water to pass. To this day, the use of dispersants to break up the oil slick remains highly controversial. I had worked success-fully with, and admired, former EPA director Carol Browner on Ever-glades restoration. But I was shocked to hear her claim on the *Today Show* that the oil was essentially "gone." In fact, the dispersants served only to make the oil disappear from where we could see it. The oil globules even-tually sank, force-feeding the Gulf the mother lode of oil, but out of sight and out of mind. Subsequent ROV and submersible expeditions would show delicate deepwater corals smothered with the oil—dead.

As horrific as this spectacle was, it was not until I had a chance to sit behind my laptop that I was faced with something that I found more chilling than anything I had encountered thus far. Along with thousands of others, I would regularly visit the website of ROFFS, a Florida-based consulting company involved with fisheries oceanography, environmental science, and satellite remote sensing. Their bread and butter is to provide fishermen with analyses and forecasts for fishing conditions in the Gulf of Mexico, but as a public service, they offered some of the best data—both real time and modeling predictions—for the spread of the BP spill. The latest prediction showed that a portion of the oil, already being swept rapidly to the south, would impact Cuba's northwestern coast. I had pre-viously dived the area in collaborative research with our Cuban colleagues. It was beautiful, rich with healthy coral reefs.

Alarmed, as soon as I returned to DC I called an all-hands meet-ing, including members of the State Department, NOAA, a U.S. Coast Guard official based at the U.S. Interests Section (embassy) in Havana, and several others. I briefed them on the situation and the potential loss of some of the healthiest coral reef ecosystems in the Gulf, not to men-tion the political implications. The emergency meeting was not without a profound and rather embarrassing irony. The United States had been expressing concern that Cuba, with one of the largest oil platforms in the world, had been prospecting for oil off its northern coast in waters deeper than Deepwater Horizon. Models showed that 90 percent of oil from an oil spill in those Cuban waters would be swept up in the powerful currents of the Florida Straits and wash through the Florida Keys and up Florida's east coast. The tables were now turned. The oil from Deepwater Horizon was making a beeline for Cuba, and quickly.

Though I don't recall my exact words, they were to the effect of, "Okay, folks. What's the plan?" The response was silence, and the stares I received in return confirmed my suspicion: We were caught flatfooted—there was no plan. The U.S. government's dusty Rolodex was 50 years old. There were virtually no government-to-government connections with the scientific community or other experts to deal with such an ordeal. The voice of the Coast Guard official over the speakerphone offered a glimmer of hope. There were actually some good contacts with the Cuban Coast Guard, but their focus was primarily drug interdiction and rescuing *balseros*, Cubans fleeing the country to the United States aboard rafts, crossing the perilous waters of the Florida Straits. It soon became apparent that we in the NGO community had the best contacts and would need to serve as the bridge from our government to the Cubans.

I contacted our Cuban colleagues and found them without any mechanism to communicate with our government about the spill. They had virtually no information about the spill and were shocked to hear my report of the impending disaster. I asked, "What do you need?" They asked for maps, modeling studies, and information—technical information for best practices for oil spill cleanup. It was chilling to hear that they had virtually no oil spill response experience. It fell to me to create a website, essentially a clearinghouse of information, including daily updates of the ROFFS maps and information on best practices for oil spill cleanup. But there was one more issue: The Cubans had no equipment to deal with a disaster of this magnitude. With the economic embargo in place, how could we facilitate transfer of equipment, where would it come from and who would pay for it? Clearly a bureaucratic rat's nest was unfolding while the oil crept closer and closer to Cuba's waters. One thing is certain about bureaucracy: It is not designed for speed.

The Cubans and we were becoming despondent as it became increasingly clear that we were powerless to mitigate the maelstrom of the impending black tide. Exhausted, I somehow managed to get a few hours of sleep. I awakened to a miracle. The latest ROFFS map had dramatically changed. An enormous eddy had strengthened in the western Gulf, and as it spun like a whirlpool, it drew the oil into it and away from Cuban shores. The oil would eventually sink and cause damage in the deep Gulf, but Cuba's coral reefs were safe.

But as we let out our collective sighs of relief, we were forced to face the profound consequences of the lack of diplomatic relations and government-to-government cooperation. Cuba and the United States are neighbors—we share a neighborhood. You may not like your neighbor, but when a disaster strikes, neighbors need to work together for the common good. It's why collaboration in marine science is so important. But, sadly, such an important lesson appeared lost on the U.S. side. Despite some champions within NOAA, the political result was inaction. While working on the Everglades restoration project, I had established a good relationship with then–Florida governor Jeb Bush, who proved to be a strong supporter of the project. During a visit to Tallahassee to meet with his staff and raising the issue of the need for a coordinated response to oil spills—thinking then more about Cuba's deepwater drilling activities—I encountered a room of shaking heads and shrugged shoulders. In Florida politics, Cuba was untouchable. Eventually, the report of the National Commission on the BP Deepwater Horizon Oil Spill gave lip service to the need for the United States and Cuba to coordinate efforts, but nothing ever came from it. Nor did things change on the Cuban side. And so those of us in the NGO community had to maintain our role as the de facto diplomats until diplomatic relations could one day be normalized.

You Can Look, but You Better Not Touch

I often hear that divers and snorkelers are a major detriment to the health of coral reefs. Indeed, I've seen clumsy divers and snorkelers scrape against corals, grab corals, and accidentally break corals. The damage I've seen from divers, however, has been minuscule against the backdrop of the myriad of other issues coral reefs face. As friend and colleague Dr. Sylvia Earle—renowned ocean scientist and explorer—has noted, the benefit of divers and snorkelers as advocates for healthy coral reefs surely outweighs the minimal damage they cause. That's not to say I haven't seen some horrors. Photographers (sadly, myself included) can be the worst offenders, so focused on "getting the shot" that they inadvertently float into coral formations. I remember the awkwardness I had of having to physically pull a world-renowned photographer off the coral upon which he was lying to shoot a colorful nudibranch. When visiting Tobago, I was in disbelief to hear that visitors to Buccoo Reef were given plastic "jelly shoes" so they could walk atop the reef crest. In Cuba and other places, I've pleaded with local divemasters to better educate my groups and enforce the rules underwater, trying to reassure the divemasters that the conservation-minded divers would appreciate their efforts. Sadly, this has always been difficult. Divemasters are concerned that if they are too strict, they will jeopardize their tips. Too often the responsibility falls to divers to self-enforce rules and diving etiquette. In the end, though, I'm with Sylvia. With strong education and vigilant divemasters, the damage divers might do can be more than balanced by the good they can ultimately do as advocates and ambassadors for coral reefs.

Where a Great Civilization Once Stood

Our team was giddy with anticipation, especially Dr. John "Wes" Tunnell, associate director of the Harte Research Institute for Gulf of Mexico Studies at Texas A&M University–Corpus Christi, where I was an advisory council member at its founding and established its Cuba program. This wasn't a journey to Cuba, however—we were headed, in 2002, to Veracrúz, Mexico, aboard a Mexican navy vessel to explore the stunning, renowned reefs there that Wes had studied a decade earlier. An expedition of National Geographic Society's "Sustainable Seas Expeditions," we used two manned submersibles to explore the massive reef system.

Seated inside the *DeepRover* submersible with great anticipation for a vibrant reef that lay below me, I was lowered from the deck of the ship and released into the warm blue waters. I radioed the ship that I was going to begin my descent. As the reef came into view, my eyes seemed to betray me. The rich colors of the coral reef were absent. So were the fish. Where a once-magnificent coral reef had stood less than a decade earlier, only a massive skeleton remained, now covered with algae and little else. Flying the sub between massive canyon walls built over millennia by once-thriving corals, the scene was reminiscent of World War II newsreels depicting Europe's bombed-out cities, their former glory evidenced only by the lifeless shells of long-abandoned buildings. I saw only four large fish during my six-hour dive.

Veracrúz is yet another textbook example of what happens when a lethal mix of the many factors impacting coral reefs is orchestrated into a multifront assault. Along an 11-kilometer stretch, 57 pipes spew untreated sewage into the coastal waters, the aroma quite noticeable walking along the promenade at water's edge. Obviously, the area has been robbed of its large fish. But these reefs had another factor working against them, originating from 100 miles away: deforestation. Without the root systems to hold the soil in place, it washes into the streams and rivers and makes the long journey all the way to the reef. In addition to the algae, a layer of sediment was also clearly visible.

In 2003, a cover story in the journal *Science* reported a "massive region-wide decline of corals across the entire Caribbean basin, with the average hard coral cover on reefs being reduced by 80%, from about 50% to 10% cover, in three decades." The article's prognosis is bleak. "The ability of Caribbean coral reefs to cope with future local and global environmental change may be irretrievably compromised." The precise estimates of coral cover have varied since the article was published, some studies revising the 80 percent figure downward to 50 percent, but wherever the exact number may lie within that range, it's clear that the future of coral reefs in the Caribbean lies in the balance.

It's depressing. When giving a speech, it's usually at this point that I apologize to my audience for depressing the hell out of them. But I also promise to give them hope, with an image of a DeLorean, the modern Hollywood icon of a time machine. I then reveal my next slide. Replacing the DeLorean is a 1952 Dodge on the streets of Havana. At a time when many of us were giving up hope, Cuba was the time machine of hope we had been searching for.

CHAPTER SIX

OMG, I THOUGHT YOU WERE DEAD

We will do what we have always done. We will find hope in the impossible.

— Mr. Spock

Paraíso

You've seen it in the faces of infants when they recognize their mother's smiling face above the cradle of her arms. You've seen it on the face of an old friend across the room when she suddenly recognizes you . . . after all those years. And Doug Shulz, producer at Partisan Pictures, saw it clearly on my face after he tapped me on the shoulder and pointed toward an old friend I hadn't seen in more than three decades.

When we humans recognize a friend, our faces convey it with a distinctive widening of the eyes. Combine that with the surprise of seeing someone we aren't expecting to see and our eyes grow even wider, often accompanied by a cartoon-like jaw drop. Judging from Doug's expression while observing my face, I can only imagine how wide my eyes were. Since we were 20 feet beneath Cuba's Gulf of Mexico waters, it must have been difficult for him to discern between an expression of surprise and delight versus a textbook example of wide-eyed diver panic. My eyes were transfixed on my old friend with a funny name whom I hadn't laid eyes on since I was a teenager. Larger than life, vibrant, and embracing the sun, my friend was very much alive and healthy, clearly enjoying the good life in Cuba.

At times I had doubted I would ever see a large, mature stand of elkhorn coral again. I had seen small fledglings in the Virgin Islands and elsewhere in the Caribbean, and more mature, healthy stands during an earlier 2004 expedition in Cuba's Gulf of Mexico waters, but here stood an enormous formation, flaunting itself as dramatically and triumphantly as it had a generation ago, before most of its kind vanished from the Caribbean. Elkhorn coral has been described as the "poster child" for coral decline in the Caribbean, decimated by bleaching, white band disease, hurricanes, algae overgrowth, and the rest of the long list of things that corals hate. A number of scientific papers, pointing to the nearly 95 percent loss of this coral and its cousin, staghorn coral, in areas like the Florida Keys, have noted that such a grave loss has seriously altered "the fundamental dynamics of shallow-water community structure." So emblematic is elkhorn coral of the healthy coral reef, and so heart-wrenching has been its loss, that, while vice president at Ocean Conservancy, I lobbied hard—and won—to have its image included in the organization's redesigned logo. You can't miss it, at the bottom, to the left of the humpback whale.

We were near Cayo Levisa, on Cuba's northwestern coast. On our journey the previous day from Marina Hemingway just outside Havana, I had taken the helm on the flying bridge of our rented cabin cruiser from our captain, Rolando, navigating with my iPhone. (The vessel wasn't equipped with GPS navigation.) While the others were below, I watched the sun sinking toward the horizon. I was savoring this peaceful, tranquil moment when I realized that we had motored for more than three hours without seeing a single boat. In Florida I'd be a nervous wreck trying to avoid swarms of boats on the water.

I knew from data and old grainy photos taken by colleagues that corals flourished in Cuba. And on previous expeditions, I had even glimpsed tiny patches of medium-sized elkhorn, clinging to reef crests, standing tall before the breaking turquoise waves. But in my wide-eyed encounter, I was breathless before this giant. And I beheld not just a small patch of healthy coral. I saw stand after stand—a forest of glorious, healthy mustard-brown *Acropora palmata*, as far as my eyes could see in the fading afternoon sun. Doug and renowned cinematographer Shane Moore had found it before I spotted it and were already capturing frame after frame of video for the PBS series *Nature*, for an episode that would be called "Cuba: The Accidental Eden," and would premiere as the leadoff season episode in 2011. As they filmed, all I could do was float and stare . . . and occasionally breathe.

During those years, I was overwhelmed by a whack-a-mole game we were losing, the seemingly insurmountable task of simultaneously dealing with an impossibly long list of coral-killing impacts. Protecting—let alone restoring—coral reefs was beginning to feel impossible. My hope was beginning to give way to despair, nearly abandoning any confidence in a future with coral reefs. It was Cuba and its remarkably healthy reefs, like this one, that resuscitated my waning hope.

What made this sight even more incredible is what we had just seen above the surface. Nearly a year ago to the day, not one, but two major hurricanes—Gustav and Ike—converged on this area within a week of each other, causing tremendous damage. The storms tore millions of leaves from the islands' protective mangroves, leaving a tangled fringe of rotting brown branches along the coastline. What were formerly aids to navigation were now, as Shane pointed out, hazards to navigation, bare wooden posts protruding from the channel, stripped by the winds of their painted markers and lighted beacons. And there was damage underwater, too. The storms toppled dozens of corals, especially elkhorn, which lay on their sides or were broken into small piles of coral rubble. Some of them were massive, surely a century or two old. But even among such wreckage there was cause for joy. Already the elkhorn corals were growing back, and rapidly so. Many of the dark-brown algae-covered dead branches were tipped with bright mustard and white extensions several inches long— healthy, young coral exhibiting a quality that conservation biologists long to see in organisms like corals: resilience, the ability of species to rebound from untold stress, to endure while others perish.

As we rested aboard our boat between dives, a lobster fisherman paddled into view in a tiny *pneumatico*, essentially a truck tire inner tube. The sight of this lone fisherman, gliding upon the warm emerald waters against the backdrop of this small, tranquil key and the dramatic *mogotes*—flat-topped mountains—of Pinar del Río Province along the distant mainland, it seemed a scene conjured up by the pen of Ernest Hemingway. As if reading my mind, our captain and guide, Rolando, pointed toward the key and identified it as Cayo Paraíso, Paradise Key, so-named by Ernest Hemingway himself. It's not the official name of the tiny island, but the locals and the nautical charts all refer to it as Cayo Paraíso. The tiny crescent-shaped island, fringed by mangroves along its southern shore, was Hemingway's Cuban hideaway during and after World War II. During the war, he patrolled the waters along Cuba's northern coast aboard his yacht, *Pilar*, in search of German U-boats, which were inflicting a dreadful toll on merchant shipping in the Gulf and Caribbean. His experiences would be the inspiration for his 1970 novel, *Islands in the Stream.*

Rolando reminisced about camping on Cayo Paraíso with his father. The hurricanes of the past year had washed away nearly half of the island, and I detected a flicker of sadness on Rolando's face. But at the same time, I couldn't help but think about the *paraíso* I had just seen underwater. Certainly, it was one of a handful of places in the Caribbean that still looks as it did when Hemingway plied these waters. I imagined him returning, wide-eyed, to greet his old friends.

Even before I had a chance to explore the coral reefs on Cuba's southern coast, which would be even more remarkable, the sight of these healthy, mature corals was a gift of hope—and one of curiosity. Why do corals here flourish while just 90 miles to the north in the Florida Keys, and throughout the Caribbean, they lie dead and dying? Is this indeed an "accidental Eden," an accident of history due to the fact that Cuba took a dramatically different socioeconomic path, or is there more to the story? Deciphering this mystery has been central among the goals of our ongoing collaborative research efforts with our Cuban colleagues, and to this day we continue to piece together the answers.

ACCIDENT OF HISTORY

For life and death are one, even as the river and the sea are one.

—Khalil Gibran

CHAPTER SEVEN

MAY I ORDER BEEF?

Temper us in fire, and we grow stronger. When we suffer, we survive.

—Cassandra Clare, *City of Heavenly Fire*

Because There Is None

"May I order beef?" asked Dr. Ana María Suarez, vice director of CIM.

I was confused by the question. Ana María was one of Cuba's most respected scientists and leaders, and it was nothing less than awkward that she would ask my permission to order beef—let alone anything else she wanted. I turned to Ana María. "Of course you can order beef. Why do you ask?"

"*Por que no hay!*" she replied, her tone one of startling exasperation. I realized immediately that my question was terribly ignorant. "Because there is none!" she had said.

It was my first visit to Cuba. CIM's senior team and I had just sat down in one of Cuba's state restaurants where I planned to treat them to lunch. Such a meal would cost close to a week's salary, even on the senior scientists' relatively generous salaries, equivalent to around $20 per month at the time.

I knew too little about Cuba and its history when I stepped off the plane in Havana for my first visit in September of 2000. I was hopelessly unaware of what the Cubans had suffered over the past decade and were just now slowly emerging from. It was a decade not only without beef but also without oil, without electricity, and without countless other essentials of life. It was known by the mother of all euphemisms as Cuba's "Special Period." Soon enough I would learn how these years shaped Cuban society, the Cuban psyche, Cuban coral reefs, and the work that I would do for the next two decades.

A Distant Wall

When the images of the events of November 9, 1989, reached Havana, they were met by the Cuban people with disbelief, fascination, and wonder—wonder of what this portended for the future of their island. What little news there was of the crumbling of the Berlin Wall was difficult to absorb. Some quietly celebrated, nurturing a glimmer of hope that Cuba might also follow a different path that could lead to a less austere and controlled existence. Others were indifferent to events happening so far away and disconnected from the island. But those with a global and long view understood the significance of this momentous chapter of history to the empire that was the lifeblood of Cuba's economy: the Soviet Union. The USSR was Cuba's economic life support system for decades, filling a void long abandoned by the United States after the Cuban Revolution in 1959, punctuated by the U.S. imposition of a crushing economic embargo on Cuba by President John F. Kennedy in 1962 that still remains in effect to this day. If Eastern Germany could dissolve, could not the same fate await the Soviet Union, and could Cuba survive the loss of its essential trading partner? It would take only two short years to find out.

When I visited the Soviet Union as part of an environmental delegation just 90 days before its ultimate demise at the close of 1991, it was clear that the legendary tales we in the West were told of empty shelves and around-the-block lines of desperate citizens awaiting whatever lay at the front of the line (often they did not know) were not only true but worse than we could imagine. As we stared through the foggy windows of our bus during the ride from Sheremetyevo International Airport to our hotel, hundreds upon hundreds of Soviet citizens stood in impossibly long and winding queues, enduring an icy-cold gray mist in a city devoid of color for whatever morsel, if any, might await them hours from now at the front of the line.

Our interpreter, Svetlana (who eventually became my wife), thin and hungry, remarked that the meals she had with our delegation were the best she'd had in years if not her entire life. To her, it was especially unimaginable to hear plump members of our delegation complaining about the food at time when a full day of foraging for her daughter and herself would sometimes yield only a single weeks-old onion. The ample meals on our plates were, to Svetlana and others, miraculous, painstakingly obtained through the black market after intense negotiations and numerous bribes. And our Soviet hosts were sure to ply our group with vodka and cognac throughout the night as delegation members tried in vain to keep up with the limitless alcohol capacity of their Russian counterparts.

Already the Soviet Union could barely feed itself, and its support of Cuba was stretched to the limit. Three months later, the lowering of the Soviet flag from the mast atop the Kremlin and the raising of the tricolor Russian flag signaled the end of the decades-old support for Cuba. The newly emerged Russian Federation swiftly abandoned the shipments of petroleum and other aid that had been guaranteed to Cuba, and the island suddenly found itself alone and untethered. Some wondered whether this spelled the end of the Cuban Revolution. It came to represent the island's most daunting postrevolutionary challenge.

An Island Hungry and Adrift

The collapse of the Soviet Union brought Cuba to its economic knees. Without Soviet aid, economic assistance vanished, as did food, oil, building supplies, and a sweeping range of essential items that kept Cuba's Soviet-addicted economy afloat. In just four years, Cuba's gross domestic product decreased by 35 percent. Construction fell by 74 percent; food production by 47 percent; and manufacturing capacity by 90 percent, largely due to vanished oil imports from the Eastern Bloc. Cuba is considered by many to be one of the only countries to endure and survive "peak oil," the point at which oil enters an irreversible decline.

Without oil, electricity was available for only a few hours each day. At night, Havana was cast into darkness. Gaspar González, in his typically gregarious and colorful fashion, recalled those nights and how he—ever the scientist—used the opportunity of a pitch-black city to bring his two sons to the roof of their apartment building and teach them astronomy beneath a brilliant bounty of stars, blanketing the city with color and detail his young sons had never beheld.

The chronic petroleum shortage transformed Cuba's transportation system as bicycles and horse-drawn carriages replaced cars and buses. Few Cubans own cars, even today, and with hours-long waits for buses and other public transportation, hitchhiking was (and remains) commonplace. In their resilient humor, Cubans refer to it as traveling by *botella* or bottle, explained to me by a friend that the outstretched thumb from the fist—the common signal to hail a ride—resembles a bottle. For good measure he brought his thumb to his lips and tipped his fist upward just in case I didn't get the analogy.

If you're lucky, the *botella* could get you a ride in a comfortable government car. More often, however, your morning commute would be in the bed of a dump truck. Public transportation was unreliable and, concluding that public transportation vehicles needed to haul more passengers, the government retrofitted 18-wheelers with trailers designed to hold 100 passengers—and many more as Cubans crammed themselves into every available space. Known as *camellos* (camels) for the two strange humps of the trailer, a few remain in service today outside Havana. While I was having a curbside chat with colleague Dr. Sergio Pastrana, now head of the Cuban Academy of Sciences and always brimming with humor, a *camello* pulled up along the curb where we were standing. He jokingly urged me to board. "David, you not only get a ride; you also get a sauna at no extra charge!" Getting to work could take hours. In the early 2000s, it took CIM graduate student Daylin Muñoz more than two hours of hitchhiking and at least three separate vehicles to reach the center from her home in Bejucal, and, of course, another two-hour journey to return home at night. Cuba purchased and distributed 1.2 million bicycles from China and manufactured another half million itself. To this day, many Cubans who endured the Special Period eschew and avoid bicycles. To them, they remain a painful icon of their suffering during the nineties.

But it wasn't the blackouts nor the *camellos* nor the bicycles that had the greatest impact on the Cuban people. It was hunger. Food consumption dropped by up to 80 percent, and the average Cuban lost around 20 pounds. More than bicycles or anything else, it is the memory of years of hunger that Cubans carry to this day. For the average Cuban, daily life became a ruthless struggle, and it fell to the Cuban people themselves to find ways to survive. The country managed to avoid outright famine, but desperation for food led to extraordinary measures to stay fed and keep one's family nourished.

"My neighbor ate my cat," said my dear friend Vilma Albelay—a Cuban American living in Washington, DC. Her face reddened, her eyes and lips tightened. "He denied it but I *know* it was him." Decades later she was still filled with anger and grief over the loss of a beloved pet. She survived those times as a hungry teenager, but was more fortunate than most thanks to the fact that her father, Ramón, was a fisherman and was able to keep his family fed.

During the Special Period, ocean wildlife experienced new pressures as many turned to the sea for food. Fishermen, by trade or otherwise, would catch and eat not just fish and lobster, but eagerly consume virtually any sea life that was safe and marginally palatable. Sea turtles—entangled in nets or found nesting on beaches—were a delicacy, as were their eggs. My colleague and friend at CIM, Dr. Rogelio Díaz-Fernández, recalled when magnificent pink flamingos, adorning rich and verdant seagrass beds near Cayo Coco along Cuba's northern coast, became the target of hunters. They were cleaned, packaged, and sold as chicken in local markets. An enormous manatee or "sea cow" would feed an entire neighborhood with nourishing meat like that from their terrestrial cousins. If caught in a net, they would be butchered and sold. Killing a sea cow is illegal, but doesn't carry anywhere near the penalty of killing a cow on land. To this day, doing so comes with a dreadful cost. Slaughtering cattle is only allowed by the Cuban government. Anyone killing a cow—even one they raised—is subject to 10 years in prison—sometimes more. Yet there are dozens of stories of desperate Cubans sneaking into a field at night to kill and butcher a cow. Cows killed by lightning or struck by a train would trigger a race to abscond with the meat before the authorities arrived. Rumors persist to this day of animals taken from the Havana Zoo for food during the nineties.

However, the typical Cuban, without turtle, flamingo, manatee, or illegal beef on the table, had to make do with what meager portions of rice, beans, milk, eggs, and cooking oil were afforded at local state-run bodegas, strictly rationed and documented on citizens' precious government ration cards, a practice still in effect today. The rations fell well short of fending off persistent hunger, and Cubans needed to find creative ways to feed themselves. Gaspar González's wife, Dr. Consuelo "Coqui" Aguilar, also a scientist and colleague at CIM, recalled how she would meticulously pare a grapefruit's pith—the white flesh between the peel and the fruit—to make small patties. "I would put bread crumbs on them and fry them and they tasted like steak!" Coqui told me years later, kissing her fingertips to convey the wonderful flavor of the unusual creation. I could see in her eyes that the taste of the rare delicacy offering a brief respite from hunger still lingered on her tongue a decade later, along with the satisfaction of finding an innovative way to feed her husband and two growing sons.

For more than 40 years, Cuba's equivalent of Julia Child, Nitza Villapol, hosted *Cocina al Minuto*, a popular cooking show on national television. Many of my Cuban friends recall Villapol, defiant of the food shortages, teaching viewers how to prepare a Special Period–version of *ropa vieja*, a signature Cuban dish of shredded beef. To their astonishment, she substituted plantain peels, chopped and spiced, for unattainable beef in the traditional Cuban dish, extolling the nutritional value of the dish. Among my friends, none were tempted to try the recipe.

During the early years of the crisis, private U.S. groups provided humanitarian aid, primarily food and medicine, permitted under U.S. law at the time. However, in 1996, two Cessna aircraft operated by a U.S. group calling itself "Brothers to the Rescue," a Miami-based humanitarian search-and-rescue support group working to rescue rafters trying to make the treacherous journey across the Straits of Florida from Cuba to the United States, entered Cuban airspace and were shot down by the Cuban Air Force, triggering outrage in the South Florida Cuban American Community and Congress, leading to tightened regulations and making it much more difficult to continue humanitarian support.

As much as Cuba's hunger grew, its ability to feed itself diminished. Broken-down Russian tractors littered the countryside, frozen in their tracks, awaiting parts that would never arrive and fuel that would not flow for years. The Soviet system of centralized and industrialized agriculture collapsed and rusted in the tropical humidity, the country's transportation system now incapable of distributing what limited harvest there was before it would spoil.

It was not unlike what happened in the Soviet Union itself. My wife, Svetlana, born in Moscow, recalls the mandatory duty of every Soviet college student to spend weeks away from home in distant collective farms, digging potatoes from hard, frozen soil, potatoes either frozen solid or soft to the touch with black rot. As if the backbreaking work wasn't hard enough, the female students had the additional burden of fending off members of the Uzbek army platoon assigned to oversee the students. "They were grabbing our asses" as the young women were bent over digging up the potatoes, recalls Svetlana. In Moscow, her duties also involved working from midnight to 6:00 a.m., shivering in an unheated and dilapidated warehouse, sorting cabbage into two piles: the deceased and the undead. The latter, marginally edible, still had miles to travel before they reached the hands of hungry Soviet citizens who were waiting hours in line to purchase a single head—the legal limit—of rotting cabbage. While the students pawed through the frozen-solid cabbage, none of which had a "hint of green," Svetlana observed, the staff huddled in a heated corner of the structure, passing bottles of vodka throughout the night. The stories of Soviet citizens in shock during their first visit to an American supermarket are not exaggerated. I found Svetlana standing motionless, fixated on a cabbage she cradled in her hands at a Vons supermarket in Ojai, California, during her first visit to the United States in 1992. "My God, I've never seen a cabbage so healthy."

As much as anything, Cuba's omnipresent black market was of enormous importance in sparing the country from an irreversible slide into poverty. In many places, the term "black market" is reserved for gangsters and cartels. But the face of Cuba's black market is ordinary grandmothers, children, and families. It touches just about everyone and is the grease that keeps the gears of a dysfunctional economy in motion. While I was visiting a family in Bejucal in 2003, a neighbor approached the open window of their home. After exchanging pleasantries, she asked if we needed any coffee. Small-scale theft by factory workers was commonplace, especially at food production facilities. As the neighbor walked to the house next door, my friends explained: "In Cuba, your neighbor is as important as your family," a refrain oft repeated. Cubans were in disbelief that when I lived in California, I almost never saw my next-door neighbors. I explained that as they were driving and neared their home, they would press a remote control to open the garage door. The door would raise, the garage would swallow the car, and within seconds the door was again closed. A high fence separated our properties. We never saw or spoke to one another. I never did know their names. In Cuba, your neighbor is your lifeline, whether it be for finding food or medicine, sharing information, or caring for one another in just about any aspect of life. Driven in part by necessity, Cuban communities are tight-knit, warm, and endearing. American visitors who are old enough to remember days gone by feel a bittersweet sense of nostalgia, recognizing a void today in American communities created by the viral spread of exurbs, movement toward greater self-reliance, isolation from others, and now relegating more and more of our social interactions to the virtual world of social media.

Fertilizers and pesticides supplied to Cuba by the Soviet Union vanished while what limited crops there were could barely keep up with the voracious appetite of hordes of tropical insects. If Cuba was to survive, it would need to radically change its agricultural practices. And so it did. The Special Period gave birth to a new era of agriculture in Cuba, one that has endured to the present. And it is widely believed that this change, albeit born of a dreadful, unplanned twist of history, is one of the most important explanations of why Cuba's coral reef ecosystems have escaped the ruin that has befallen these treasured ecosystems around the world.

CHAPTER EIGHT
A RIVER ON FIRE, A RIVER OF GRASS

The Everglades is a test. If we pass it, we may get to keep the planet.

—Marjorie Stoneman Douglas

The River That Sparked a Revolution

It was 11:56 a.m., just four minutes before the noon whistle would sound on a typical weekday. Workers from the Republic Steel mill on the banks of the river would be finding respite from the oppressive heat and noise to sit and enjoy their lunch, some quiet, and conversation. But this was a Sunday, tranquil and unhurried. The June morning's calm surrendered only briefly to the occasional rattle of a Norfolk and Western diesel locomotive lazily pulling its load across the river atop the railroad trestle adjacent to the mill. Such crossings were routine and practically unnoticeable in the heavily industrialized Cleveland of 1969 as it would have been for this particular train if not for the briefest contact of wheel against track, throwing a spark into the air that slowly wafted down to an awaiting five-square-foot oil slick atop the Cuayahoga River, setting it ablaze.

The train employees tried in vain to extinguish it before it mushroomed into a major fire that burned the timber of a pair of train bridges and caused tracks to buckle. Frantic, a train employee phoned the fire department. Seven companies responded with five engines, a ladder truck, and other equipment as the fire quickly spread to an area of roughly 90,000 square feet before it was finally brought under control.

71

The wayward spark ignited more than the Cuyahoga River. It set off an inferno of media attention that swept the nation with horrific images of the towering flames—and the perplexing reality of a river on fire. *Time* magazine published the images, describing the Cuyahoga, brimming with industrial waste, as a river that "oozes rather than flows." The article went on to recite a joking mantra of Cleveland citizens at the time: "Anyone who falls into the Cuyahoga does not drown—he decays."

So impactful was the spectacle and so strong the national outrage, that the event set into motion an environmental revolution that would lead directly to a body of environmental legislation that would soon unfold in the Nixon administration and that to this day remains unprecedented. The bipartisan-supported legislation included the Clean Water Act, which arguably has had one of the greatest impacts on freeing "oozing" rivers from the pipes of untreated industrial waste they had been forced to swallow.

Ironically, this was not the first time the Cuyahoga had been set ablaze. It had caught fire at least a dozen times previously, dating back a century. Decades later, the *Washington Post* reported that this latest in the series was so small, brief, and unremarkable that it "barely made headlines in the local papers." It was revealed years later that now-iconic photos published by *Time* were not even of the 1969 fire but of a much larger and far more spectacular fire that burned on the Cuyahoga in 1952, 17 years earlier. The result of that editorial decision is immeasurable as the image evoked outrage . . . and the soon-to-be-realized sea change in environmental policy.

Modest efforts were under way to reduce the dumping of industrial waste, much of it illegal, into the nation's rivers, including the Cuyahoga. The *Washington Post* reported that, rather than a symbol of the declining conditions of rivers across the country, the 1969 fire "was not the first time an industrial river in the United States had caught on fire, but the last."

But while the Cuyahoga may have indeed represented the end of the era of burning rivers, it remained a symbol of a more ominous problem that lay below the flames above. The river was virtually lifeless, its inhabitants unable to endure the soup of toxic waste that washed by daily, affecting everything from fish to the invertebrates trying to scratch out a living in the muddy bottom sediments as bluntly observed by federal regulators: "The lower Cuyahoga has no visible life, not even low forms such as leeches and sludge worms that usually thrive on wastes."

Across the nation, dozens of rivers were suffering from the same unsettling malady, from the Potomac to the Missouri, to Philadelphia's Schuylkill (SKOO-kill) and beyond, brimming and stinking of the waste of industrial plants, meatpacking plants, and untreated sewage.

Lake Erie itself received so much municipal waste and agricultural runoff that it was projected to become biologically dead in a few years. Unchecked water pollution in inland waterways accounted for record fish kills; for example, some 26 million fish died because of the contamination of Lake Thonotosassa in Florida. Industry discharged mercury into the Detroit River at a rate of between 10 and 20 pounds per day, causing in-stream water to exceed the Public Health Service limit for mercury six times over. Waterways in many cities across the country served as nothing more than sewage receptacles for industrial and municipal waste.

The rate of loss of wetlands—often described as a beating heart for marine and bird populations—grew to approximately 450,000 acres per year from the 1950s to the 1970s. Today, on average, most states have lost half their wetlands, California more than 90 percent.

One of the most egregious cases of toxic contamination of U.S. waters defies belief. From 1947 to at least 1961, Southern California's Montrose Chemical Corporation dumped into the ocean waters near Santa Catalina Island up to half a million barrels of the now-banned pesticide DDT (dichlorodiphenyltrichloroethane), an icon of the era of toxic waste dumping and forewarned by Rachel Carson's bible of the modern environmental movement, *Silent Spring*. Montrose was also known to have regularly poured DDT down the drain where it ended up in Santa Monica Bay. DDT is a highly toxic carcinogenic and pesticide linked to a wide variety of health problems in humans and wildlife, most infamously resulting in the thinning of egg shells of birds so that they would crush when sat upon by the female to incubate the eggs. As DDT travels up the food chain, it concentrates through a process known as bioaccumulation. Worse yet, DDT persists in the ecosystem for decades, all the while wreaking havoc.

While working at EcoAnalysis, an environmental consulting firm I cofounded in California, our team analyzed data from the dump site and surrounding areas in Santa Monica Bay. Even in the late eighties and early nineties it was clear from our analysis that the lethal impacts of DDT dumped decades earlier showed few signs of waning. Every time a storm struck those coastal waters and disturbed the bottom sediments, the DDT hiding there would be released, killing off large swaths of the bottom communities living there. To this day, this tragic legacy persists. More than 70 years after Montrose dumped its first drop of DDT into the Pacific, the Center for Biological Diversity filed a lawsuit against Montrose and its successor parent company, Bayer, Inc., in May 2021.

Nixon, the Tree Hugger

When I give public talks or teach students, one of the first questions I ask is, "What do you think represents the biggest threat to the oceans?" The response instantly volleyed back is almost invariably "pollution," the iconic image of a pipe spewing Day-Glo-green toxic waste into streams and coastal waters immediately coming to mind, even among young students born decades after such scenes existed. (Pollution in the form of plastics in the ocean, which have gained great notoriety in recent years, is also a growing response.) And while the bizarre accounts of rivers on fire more than half a century ago may seem a relic of the past, concern about toxic pollutants strongly endures.

So it's a surprise when I reveal the fact that, thanks to the shock wave of the burning Cuyahoga, the resulting renaissance of environmental regulations during the Nixon administration has, over the intervening years, largely succeeded in eliminating such industrial waste. Among a dizzying suite of environmental legislation during the early seventies that included the creation of the Environmental Protection Agency (EPA) and the National Oceanic and Atmospheric Administration (NOAA), was the 1972 Clean Water Act, still considered one of the most important pieces of legislation ever implemented. Notably, in today's political environment, it's all but inconceivable to imagine a Republican president championing a sweeping environmental movement, let alone achieving support from across the aisle. But Richard Nixon helped lead the way. It speaks to the gravity of the environmental issues the nation was facing at that time, and, of course, a much different political climate.

The Act sought "to restore and maintain the chemical, physical, and biological integrity of the nation's waters," with the goals to make all U.S. waters fishable and swimmable by 1983; to have zero water pollution discharge by 1985; and to prohibit discharge of lethal amounts of toxic pollutants. And while the Clean Water Act probably can't be credited with the end of the era of river fires, which had already begun to disappear, it had teeth that compelled municipalities to ensure the quality of their waterways. The legislation even extended to destruction of wetlands. Strict limits were placed on what industry and sewage plants could discharge. Anything more would require a permit. The Act demanded oil spill prevention and response protocols and extended further to radioactive waste, medical waste, and sewage sludge left behind after clean water leaves a water treatment facility. Such sludge can be laden with heavy metals, like mercury. Overall, the Clean Water Act has received high marks—it's been estimated that each year it prevents 700 billion pounds of toxic pollutants from being dumped into rivers and streams. The rate of wetlands loss has dropped dramatically compared to the pre–Clean Water Act era.

By 1989, the Cuyahoga was not pristine, but it was "fireproof," according to the *New York Times*. Some signs of life had reappeared, including insects and mollusks. And Cleveland's water pollution control commissioner averred that the Cuyahoga no longer oozed. It "often gleams and sparkles," the *Times* reported, almost like, well, a river.

Philadelphia's Schuylkill River, famous for winding by the Philadelphia Museum of Art where Rocky Balboa triumphantly summited its steps and leaped with fists in the air in the 1976 film *Rocky*, was infamous during those years for the pungent, lethal waste it delivered to the Delaware River, so bad that locals called it the "Sure Kill" River (a moniker also given to the harrowing, accident-prone expressway that winds along the river's west bank). So it has been with more than a modicum of amazement to return to the city of my birth and find the once noxious Schuylkill choked on weekends, not by plumes of chemical discharge, but by paddleboards, kayaks, and racing sculls. The river's fish and birds have returned. And it's not stinky anymore.

Amber Waves of Green

Since 1948, radio station KBMW has been serving as the "Voice of the Southern Red River Valley," a tristate area including North Dakota, South Dakota, and Minnesota, boasting some of the "richest farmland in the United States." So why did they want to interview me, a city boy who lives for salt water? To update their listeners on the BP oil spill disaster in the Gulf of Mexico ongoing at the time, and, most important, to tell their listeners how they could help. Like many, they felt a deep emotional connection to the Gulf, even from more than 1,200 miles from water's edge, and the daily images of oil erupting from the BP well was painful to watch. They felt powerless to help. Truth is, KBMW's listeners are connected by more than emotion to the Gulf of Mexico and can essentially help the Gulf of Mexico by changing the way they do their gardening.

"A smooth, closely shaven surface of grass is by far the most essential element of beauty on the grounds of a suburban house," said landscape designer Frank J. Scott in the 1870 book *The Art of Beautifying Suburban Home Grounds*. It was during that time that the American obsession with a green lawn took root with an eye to Europe's long-standing practice. Nearly two centuries later, the words of Frank J. Scott are still firmly rooted in the American psyche: The lawn has become a social barometer, signifying wealth and class, an iconic suburban manifestation of keeping up with the Joneses. "Maintaining a well-kept lawn has become almost a moral symbol signifying the decency and probity of the homeowner," said John Lacy in the *New York Times*. Even if you don't care to compete with your neighbor for the greenest lawn, in many localities you have a civic duty to maintain a well-manicured lawn or face the risk of being slapped with a fine.

Unfortunately, our well-intentioned efforts to beautify suburbia have grave consequences due to the myriad chemicals we use to keep our lawns green and weed-free. It may seem that our lawns couldn't really amount to much as compared to enormous industrial farms and hog feedlots that spill fertilizer-laden water into the waterways. But, in fact, it is shocking to consider that the largest irrigated crop in the United States is not corn, wheat, barley, or soy. Some 40 million acres of America are covered in lawns, making turfgrass the largest irrigated crop in the country. Americans pour as much as 238 gallons of water per person per day onto lawns during the growing season. Each year we use tens of millions of pounds of fertilizers, insecticides, and weed killers on our lawns, not to mention the 800 million gallons of gasoline we burn in lawn mowers each year. Outdoor watering accounts for *more than half* of municipal water use in most areas, and homeowners often apply fertilizers and pesticides to their lawns at many times the recommended levels.

Near the end of my tenure in 2000 as president of the Conservancy of Southwest Florida and cochair of the Everglades Coalition, I planned to launch a new campaign, which I dubbed the playful if not apt "Kill Your Lawn Campaign." I planned to take a shovel to the lawn of my home in Naples and replant most of it using native vegetation, guided by the Florida Yards & Neighborhoods program, which provides training and advice on how to go about replanting with native vegetation that requires little or no watering, fertilizer, or pest control. The result is a lush, natural landscape that attracts wildlife, including butterflies and birds. And for the local water bodies, including the Gulf of Mexico, my yard would no longer be contributing to the nutrient load. In Naples and most of South Florida, there was a senseless irony that we were part of an ecosystem where water conservation is a critical part of protecting and restoring the Everglades, but we were pumping the Everglades dry to irrigate lawns, street medians, and golf courses. So, in fact, there was something that those tuning into KBMW could do to help the Gulf of Mexico, I told the interviewer. They were connected to the Gulf not just by emotion, but by water. Reducing their use of fertilizers and avoiding overwatering would make a difference. And for those ready to make the ultimate sacrifice, they could kill their lawn entirely.

Successful as Nixon's sweeping environmental regulations were in capping the pipes spewing toxic waste, so-called point sources of water pollution, it became increasingly clear in the years following that the Clean Water Act was falling short. It wasn't that the regulations were weak—they were just the opposite. But the authors of the Act were focused on toxic waste from pipes, not lawns or farms. As the effluent from pipes became cleaner and cleaner, something else more insidious was at work, something at an unimaginably grand scale. This pollution ravaging our waterways, bays, estuaries, and coastal waters wasn't coming from pipes and other discrete point sources. It was now the polluted water flowing from everywhere else: in the parlance of scientists and regulators, nonpoint-source pollution. While point-source pollution is relatively easy to identify and confined to a particular area, nonpoint-source pollution emanates from the lawns of suburbia and the farms of the heartland. In 2009, more than 40 percent of rivers and streams had excess nutrients, a direct result of the nitrogen and phosphorous of fertilizers and pesticides washing into nearby streams and rivers on their journeys to the bays, estuaries, and ocean waters along U.S. coasts.

Amber Waves of Grain

Significant as they are, our backyards play a supporting role to the mother lode of nitrogen-based fertilizers: the amber waves of grain of the American heartland. Along with chemicals from our lawns, draining into the Mississippi River and, in turn, into the Gulf of Mexico is a massive load of fertilizers and pesticides from hundreds of thousands of acres of farmed land—much of it from sprawling industrial farms—stretching north across the entire width of the lower 48, as far west as Idaho, as far east as New York. Waters even sneak into the Gulf from north of the U.S. border, from Canada's Alberta and Saskatchewan Provinces. Approximately 44 percent of the continental United States drains into the Gulf of Mexico. Water from 31 states drains into the Mississippi, creating a drainage basin of a staggering 1,245,000 square miles in size.

The resulting fire hose of "nutrient pollution" blasting into Gulf waters has had stunning consequences. Those fertilizers ably perform the same task in the waters of the Gulf that they're asked to do on a field of corn. They promote the growth of plants and algae—dramatically so. Massive algae blooms cover the Gulf waters for miles beyond the Mississippi Delta. Once these algae die and sink below the surface, bacteria go to work on them, decomposing the dead algae, a process that requires oxygen. With so much algae to munch on, the bacteria leave Gulf waters anoxic, that is, practically devoid of oxygen. This means that anything that needs oxygen to survive underwater—like fish, mollusks, and crustaceans—are out of luck. A fish's gills—exquisitely designed to pull precious oxygen from the water—are ultimately useless in an anoxic ocean.

The assault of nutrient pollution on the Gulf and other waters is happening at a scale difficult to comprehend, with sweeping impacts on marine ecosystems across vast distances. Each year a "dead zone" forms at the mouth of the Mississippi, growing as large as the state of New Jersey. If New Jersey itself were a dead zone—devoid of songbirds, trees, deer and other wildlife—the result would be an enormous public outcry and swift action to restore it. (And despite the disparagement that New Jersey too often endures, it is far from a dead zone, harboring forests, the unique Pine Barrens ecosystem, rich coastal waters and marshlands, and countless migratory birds. New Jersey has some of the best bird-watching on the planet, drawing visitors from around the world. And, of course, its pizza is legendary.) Yet New Jersey's marine twin lies underwater, keeping one of the planet's largest and most tragic environmental disasters hidden from view and consequently little known by the public. It is mostly fishermen, regulatory agencies, and environmentalists who understand the consequences of the daily diet of nutrients the Gulf is forced to eat. The rest of us are blissfully unaware of this massive zone of death where next to nothing survives.

Trying to fix the dead zone is an effort that presents an enormous challenge that encompasses tens of millions of acres of land and crosses every imaginable type of jurisdictional border of nearly two-thirds of the lower 48 U.S. states, not to mention the two Canadian provinces. And it draws focus on how we use our lands, what we grow, what we put on the land, and what runs off. The bitter lesson it has taken too long to learn around the world is that land and water are inextricably linked. The choices we make on land have consequences in the deep.

Chicken Shit

Kayaking the Anacostia River—once reviled as a toxic, industrial tributary to the Potomac—was, to understate it, a dramatic and glorious surprise. As the DC Metro sped by on the bridge above us, the banks of this urban river—hardly a mile from the Capitol dome—was crowded with magnificent great blue herons and dozens of other wading birds. Bald eagles soared above. Their presence signaled the unmistakable fact that the Anacostia was delivering plenty for them to eat. Downstream, the Potomac empties its contents into Chesapeake Bay, the largest estuary in the United States.

Like the Mississippi, the Chesapeake Bay watershed is blind to state boundaries, drawing its waters from six states (Delaware, Maryland, New York, Pennsylvania, Virginia, West Virginia) and the District of Columbia, covering 64,000 miles and home to more than 18 million people. The bay is another vivid illustration of how land and water are inextricably linked. The land-to-water ratio of the Chesapeake Bay is 14:1—the largest of any coastal water body in the world, magnifying the importance of how what we do on land impacts what happens in the bay. A reminder is affixed to storm drains on streets throughout DC. Accompanying the image of a fish are the words "Drains to Chesapeake Bay."

One of the Chesapeake's challenges is chicken farms—lots of them. What you and I might instinctively call "chicken shit" has been re-branded by regulators with the more socially acceptable yet comically understated euphemism "poultry litter." Despite its more polite moniker, it still smells like shit. If you live on the East Coast and have a hankering for broiler chickens, there's a good chance yours came from the shores of the Chesapeake. In Wicomico County alone, poultry production makes up 78 percent of the area's agricultural sales with the production of more than 10 million broiler chickens in 2017. And those 10 million chickens poop up a storm—750,000 tons in the Delmarva Peninsula per year. The majority of this "poultry litter" is used as fertilizer in other farming operations. Unfortunately, farms aren't doing enough to keep it where it belongs. Rainstorms wash this nutrient-rich litter directly into the Chesapeake. Like the Mississippi, it's an example of how the Clean Water Act of 1972 and its focus on point sources fails the poop test. Regulatory agencies don't consider chicken farms as "Concentrated Animal Feeding Operations" (CAFOs), which are treated as point sources, as the term normally applies to pig farms and the like. Chicken farms fall into a different category and therefore don't require the discharge permits that they otherwise would. In the end, it's a massive nonpoint source nightmare of nutrient pollution, spawning algal blooms, and dead zones, threatening the breathtaking marine wildlife of the bay.

Coordinating the actions of six states to restore the Chesapeake Bay has proved an epic challenge, but with some notable success. By comparison, coordinating the actions of 31 states (and two Canadian provinces) in the Mississippi watershed would seem hopeless. It isn't, but progress has been slow. So it would reason that if there are such challenges contained within just *one state*, fixing the problem would be a breeze, right?

River of Grass

Indifference, fear, and outright hatred of the Florida Everglades have a long history. During the Second Seminole War from 1835 to 1842, army surgeon Jacob Motte famously said of the Everglades, "It is in fact a most hideous region to live in, a perfect paradise for Indians, alligators, serpents, frogs, and every other kind of loathsome reptile." He went on to describe it as a "sloppy Kansas." Echoes of this sentiment linger in the 21st century, many finding more fear than love in their heart for what they perceive as a forbidding, dangerous place far removed from the comfort of strip malls and Taco Bells. Better to pile the kids into the minivan and visit Yosemite.

Historically, the dreaded subtropical "swamp," receiving 60 inches of rain per year, was mostly covered with water much of the year. In the mid-19th century, the nation's eager push for expansion led to the idea of draining the miserable, "useless" swamp in favor of promoting agriculture. In 1850, Congress passed the Swamp and Overflowed Lands Act, setting the stage for draining the wetlands and converting them to farmlands. This perception of the Everglades persisted into the 20th century. In 1904, Governor Napoleon Bonaparte (yes, that was his real name) was a big fan of draining "that abominable pestilence-ridden swamp" as he referred to it. Homesteaders began to settle the newly drained lands, es-tablish humble homes, and farm the surrounding land. Unfortunately for them, their land, which may have remained dry for several years of lower-than-average precipitation, became submerged when the rains finally came, ruining their crops and sometimes flooding them out completely.

The situation came to a head when devastating hurricanes in 1926 and 1928 caused Lake Okeechobee, the enormous freshwater lake at the center of the state lying roughly between West Palm Beach on the east coast and Ft. Myers on the west, to rapidly swell and easily breach its levees, the resulting enormous freshwater wave rampaging across hun-dreds of square miles of land, effortlessly washing away structures, crops, livestock, and thousands of human lives. The 1928 hurricane was respon-sible for 2,500 deaths. They are still considered among the most lethal of all U.S. hurricanes on record. To prevent such a disaster from occurring again, the Herbert Hoover Dike was constructed, encircling the lake.

However well intentioned, the dike had the unfortunate result of blocking the natural flow of water from the lake—a key waypoint for wa-ters flowing south—and choking the natural flow of water into the saw-grass and cypress wetlands that lie to the south. By the middle of the 20th century the number of people settling in Florida began to grow rapidly and flood control became a priority. Providing fresh water to agriculture and developing urban centers also became a priority. So in 1948, Congress approved the Central and Southern Florida Project for Flood Control and Other Purposes (C&SF), with the daunting task of dividing the waters among these groups, all the while preventing the catastrophic flooding of the past. The Everglades were divided into Water Conservation Areas to hold water during the dry season and feed the increasing number of thirsty residents arriving daily; the Everglades Agricultural Area, an enormous expanse of land dedicated primarily to growing sugarcane; and Everglades National Park, established a year earlier by President Roosevelt. Dredges covered the landscape for 35 years as the C&SF project constructed 1,000 miles of canals, hundreds of pumping stations, and levees, making it one of the largest human-managed ecosystems in the world.

"I have to say it is one of the most effective and efficient water management systems and flood management systems on the planet. It worked like gangbusters," said Shannon Estenoz, the soon-to-be-appointed assistant secretary for Fish and Wildlife and Parks at the Department of Interior. In 2020, Shannon was guest lecturing to my graduate students at Johns Hopkins.

Shannon and I met in the late nineties at a meeting of the Everglades Coalition, a group of 50 national, regional, and local environmental organizations working to protect and restore the Everglades. We met at what was the first Coalition meeting for each of us. She drew attention as the youngest in the room and it soon became apparent that she was also the sharpest. At the time she was with World Wildlife Fund and uniquely, a civil engineer, later appointed by three Florida governors to the governing board of the South Florida Water Management District. However, one of the most unique things about her is that she is not only a Florida native but also a fifth-generation native of Key West, Florida. Little did we know that within a year or so the two of us would be cochairing the Everglades Coalition during the final sprint to pass the monumental legislation to restore the Everglades.

Effective and efficient, yes, but the water management system of the Everglades was built for a very different Florida than the one that came to be. "The engineers of the 1950s in their wildest imaginations, the largest population they could imagine would come to live in South Florida was 2 million people and that's what they built the system to serve," Shannon explained. Of course, history had a different plan. During the 1950s and 1960s the Miami metropolitan area grew four times as fast as the rest of the nation and increasingly more water from the Everglades was shunted to these newly developed areas. Today the system serves nearly 9 million people, and it's expected to continue its rapid growth to at least 15 million, perhaps 20 million.

Shannon pointed out that none of present-day Miami, Ft. Lauderdale, or Palm Beach would have been possible without the water management system, a system with the monumental challenge of both delivering fresh water to human settlements and holding back that very same water to prevent flooding. The system fed Florida's economy for decades, through the end of the 20th century. Agriculture became a major part of the economy—rich soil, lots of sunshine, and lots of land. As recently as the late nineties, I recall many meetings where it was accepted as an axiom—and promoted forcefully by development interests—that the Florida economy was a three-legged stool comprised of development, agriculture, and tourism.

"Fast-forward to the 21st century and that's really on its head," Shannon explained. "When you look at modern Florida, and you think of the 21st-century Florida economy, it's really a much more urban economy, with agriculture diminishing all the time. And that is largely the reason that the infrastructure is not working for us anymore. Not only that it's serving a lot more people than it was designed to serve; it's serving a different economy than it was designed to serve. The infrastructure we're living with today doesn't reflect those things anymore, so that's what Everglades restoration is all about." Implicit in Shannon's remarks, of course, was the added fact that the infrastructure was no longer adequately serving the exquisite ecosystem, its waters increasingly and unsustainably diverted for urban and agricultural use.

Over the years, the water management system began to reveal other weaknesses. Years with heavy rainfall and large storms would fill Lake Okeechobee to the brim, threatening to overflow the dike. An engineering solution was in place: Simply pump the excess water down the Caloosahatchee River, draining in the Gulf of Mexico to the west, and down the St. Lucee canal, draining into the Atlantic to the east. And it worked . . . until it didn't.

Water quality is a big issue in the Everglades. In its natural state, the Everglades have exceptionally low nutrient levels (nitrogen and phosphorus). To make the point, speaking alongside Fred Krupp, president of Environmental Defense Fund, at a press conference, I pulled a bottle of Perrier from the shelf behind the podium and held it up. I felt the burn of Krupp's gaze and glanced over at a puzzled and not-so-slightly worried look on his face as he no doubt wondered where Guggenheim and his Perrier were headed. "If I went out to the Everglades and poured out this bottle of Perrier, I would be violating EPA nutrient standards for the Everglades," I said. I had done the math—the phosphorous levels of a bottle of Perrier are indeed well above the regulatory limits. With such low naturally occurring nutrient levels, growing sugarcane, tomatoes, or citrus requires many tons of fertilizer. And, predictably, just like effluent from the farms of the Midwest draining into the Mississippi, large quantities of that fertilizer run off into the ecosystem from the massive agricultural lands, a co-op of growers known as "Big Sugar," long-since carved into the Everglades' northeast. Not surprisingly, a major goal of restoring the Everglades is to restore its low-nutrient water quality, and buying up agriculture land for conservation remains a priority.

Unfortunately, over the decades Lake Okeechobee has filled with nutrients from farmed lands, many locked into its sediments until disturbed by a storm. As the dead zone in the Gulf of Mexico has taught us, nutrients fuel the growth of algae, and Lake Okeechobee is no different, other than the fact that the nutrient receptacle is fresh water. Regularly, the lake is covered by a surreal bright green carpet of blue-green algae and the ensuing eutrophication, just like the dead zone in the Gulf of Mexico, robs the lake of oxygen. But whereas the dead algae in the Gulf have an opportunity to disperse on ocean currents, in Lake Okeechobee they can only concentrate within the confines of its shoreline. The lake soon becomes a cauldron filled with a warm and literally black soup of dead algae and bacteria, a toxic ooze. Not only is this disastrous for the lake itself but for the coastlines as well. When the waters reach the brim of the dike, the pumps are fired up and an enormous pulse of the toxic black water snakes its way to both coasts, regularly making headlines as dead fish and empty beaches replace the spring breakers and fishing skiffs. Corals along Florida's east coast have been dying at an alarming rate, tied in large part to the nutrient pollution that bathes them. There are constant pleas from residents, local governments, tourism businesses, and recreational fisher-man along the coasts to make it stop.

Shannon paused her remarks. After a long sigh, her face divulged a deep exasperation. She shared her preoccupation with what was happen-ing in Florida at that very moment. A low-pressure system had parked itself over the state with four days of nonstop rain. The lake was filling. A release was imminent. She explained how climate change has altered South Florida's weather, which was experiencing more frequent and much wetter storms. Once again, a 20th-century system was put to the test—and was failing—in the 21st century.

If ever an ecosystem illustrates the inextricable link between land and water—both fresh and salt—it is the Everglades, dubbed the "River of Grass," the title of the 1947 book by Marjorie Stoneman Douglas, a re-vered American journalist, author, and conservationist best known for her stalwart advocacy to protect the Everglades. However, the River of Grass is unlike any other river with its slow, gradual southerly flow of water known as "sheet flow." Its shallow waters inch almost imperceptibly across a landscape that is conflicted as to whether it is a river or a landmass, not unlike the poetic description of Florida by author John Rothchild, observ-ing it as "a geologic appendage, forever amorphous, never quite earth, never quite ocean, raised up from the Atlantic and Gulf of Mexico, only to be dunked again with the melting and refreezing of the northern ice caps."

South of the black waters of Lake Okeechobee, the River of Grass begins in earnest where its shallow waters surrender to gravity and lazily edge southward, their final destination patiently awaiting along Florida's shores to the south. The estuaries into which the Everglades' waters empty are the critical nursery grounds for important species of fish, many of great commercial importance. An estuary's health depends on a very delicate balance of fresh and salt water. Development in the Everglades ecosystem and the alteration of the natural flows of water have caused un-told damage to these estuaries, and in turn, to Florida's beloved ocean life.

I recently came upon a nostalgic relic: the transcript of my testimony to Congress in 2000. I was making a pretty big ask of the Environment and Public Works Committee—actually an unprecedented ask—all the while trying to maintain my composure, keep my sweating brow in check, and accurately recite the remarks that so many of my extraordinary col-leagues had helped me write. The ask? To spend billions of taxpayer dollars over decades on the largest and most expensive environmental restoration project ever attempted in history: the restoration of the Everglades. We were overjoyed when Congress (and the Florida Legislature, which would share the cost), in a remarkable example of bipartisanship, said yes, launching the replumbing, reengineering, and reimagining of the massive ecosystem. Now more than two decades in, the project strives to restore the slow-moving sheet flow of water, the lifeblood of this unusual wetland, and ensure it is once again free of contamination and flows to the right place, in the right quantity, at the right time.

The Everglades—most commonly thought of as the national park carved from its southeast corner near Miami—are in fact so expansive that many don't realize they are living in them. The massive ecosystem stretches from the headwaters of the Kissimmee River miles north of Disney World in Orlando, south along both coasts, across Florida Bay to the Florida Keys and to the coral reefs that lie just a few miles south beneath aquamarine waters. When I began participating in public meetings about restoring the Everglades, they were exclusively in Miami, the focus of Everglades restoration for decades. I tried and ultimately, with much help, succeeded to change this bias by using the term "Greater Everglades Ecosystem" to underscore the massive, interrelated nature of this complex and remarkable ensemble of alligators, panthers, wetlands, sawgrass, moss-festooned cypress trees, standing water, manatees, wading birds, sea turtles, mangroves, and coral reefs, to name just a few residents of this phenomenal place, erupting with life in every square inch, all deeply dependent upon one another to create a living landscape like no other. It sings with life and mystery—and mosquitoes.

In the late nineties, the years leading up to the first major Everglades Restoration bill, I felt like an outsider. Southwest Florida, where I was based, seemed to be all but absent from restoration discussions. But little by little, heads turned west, shining a spotlight on this beautiful and critical southwestern corner of the ecosystem that had yet to become the miles of pavement known as Miami. Until then, residents of Naples and Ft. Myers hadn't seen themselves as part of the Everglades, which, in their mind, lay far to the east. Most were surprised to hear of their "newfound" connection to this magnificent ecosystem and proudly embraced it.

Shuttling back and forth from Naples to Washington, DC, cardboard tubes of maps tucked under my arm, I met with William L. Leary, senior counselor to the assistant secretary for Fish and Wildlife and Parks at the Department of Interior. Bill immediately got it. Over many months, he championed the cause of the Western Everglades, including the urgent need to prevent the largest subdivision in the world from being constructed. More than 78,000 acres, then known as South Golden Gate Estates, this building project was the poster child of "selling swampland in Florida." The tales are legendary of high-pressure sales of tiny lots, often sight unseen. Unsuspecting northerners toured the Estates during the dry season, unaware that the lot they would purchase in December would be underwater by June. Some prospective buyers were taken up in a Cessna, where the pilot would open his window and drop a bag of flour, marking a parcel of land. "That's your lot," he would say.

South Golden Gate Estates was a bizarre place, platted and crisscrossed with long roads that led to nowhere, devoid of structures. Rumors abounded about drug-running planes from Colombia and other Latin American countries using the roads as runways to off-load tons of cocaine to awaiting vehicles. It was even rumored to be the site of Contra training camps. But the massive land of the Estates represented an indispensable linchpin in the restoration of the natural flow of water to the estuaries, like Rookery Bay, and the beautiful 10,000 Islands that adorned Florida Bay's northwestern margin. Its development would be a devastating loss to the goal of restoring the natural hydrology and flow to the coastal waters.

At one point I insisted that one of our staff members at the Conservancy of Southwest Florida, where I served as president, Ruth Didonato, be placed on the cover of our annual report. That year, our cover shot wasn't a typical environmental organization's shot of a comely young naturalist rescuing a sea turtle. It was of a middle-aged woman, her desk groaning under the weight of countless land deeds. It was her job—working with the State of Florida—to complete the impossible task of buying back thousands of lots, a critical albeit unglamorous part of conservation that the public rarely sees. Ruth spent each day on the phone contacting the owners of record. Many had died. Many didn't know they owned the land. Many were elderly and simply confused. And to make matters worse, funding to purchase the land was running out. To make an impossible task even more difficult, local government did not support the effort. The Collier County Board of Commissioners preferred the narrative where the Everglades were indeed a distant patch of swamp, freeing them to permit golf course development after golf course development farther and farther into the Western Everglades, undermining conservation efforts of state and federal agencies and slicing the wildlands into a patchwork of homes, roads, manicured medians, and shopping centers.

The call from the White House came without warning. A fast-talking aide to Vice President Al Gore dryly informed me that the vice president would be coming to Everglades National Park and that I would be sharing the podium with the him as he announced $25 million of federal funding. The funding would be matched by the state to complete the acquisition of South Golden Gate Estates and prevent its construction. I hung up the phone, and any offense caused by the abrasive manner of the vice president's aide was quickly swept away by the euphoria in knowing what monumental news this was. And so, a few days later, on the hottest, most humid day imaginable, I took my seat between Vice President Gore and Lt. Governor Buddy McKay. The vibe from my Everglades colleagues—several inexplicably wearing suits and ties soaked through with perspiration—was unmistakable. They puzzled about this newcomer, Guggenheim, drawing attention away from the historical East Coast focus of Everglades restoration. But largely thanks to visionary Bill O'Leary, who had flown from Washington with the vice president that morning, a short but sweaty ceremony concluded with what I still consider one of the greatest accomplishments of my career—one, I must stress, that many others were part of as well. South Golden Gate Estates would be no more, replaced by Picayune Strand State Forest, now undergoing hydrological restoration as a key component of the restoration of the Everglades.

Despite the success, there was and is more to be done—much more. Exasperated scientists near Naples deeply fear for the future of many coastal waters that lie at the terminus of the Everglades southerly sheet flow, including one of the most significant estuaries in the region, Rookery Bay National Estuarine Research Reserve. With water flow into Rookery Bay severely disrupted, much of the upstream vegetation replaced by concrete, the water arrives at the bay without the buffering that cleanses and slows the water. We've turned slow water into fast water, and a single storm event can result in the bay's brackish waters surrendering to the inundation of fresh water. Such a rapid change in salinity can be lethal to many juvenile species of fish, crustaceans, and other species. It's an especially cruel irony that we've managed to make fresh water a pollutant.

I have fond memories taking donors for pontoon boat excursions upon Rookery Bay's tranquil, shimmering waters at sunset, sipping wine and gazing skyward as an endless wave of herons, egrets, and other magnificent birds, in near silhouette against a fiery sky, returned to roost for the night in the safety of the tops of the lush mangroves fringing Rookery Bay's shores. The gentle flap of each bird's wings was clearly audible, and the din of our chatty Naples crowd soon fell silent, mesmerized by the beauty of the visual and aural spectacle above us. Fish splashed. Dolphins sounded in silhouette before the growing orange sun as it made its plunge into the Gulf of Mexico. One of our naturalists, Beth Mason, turned to me during one such moment and whispered, "This is how we get paid— we call it 'psychic pay.'" She and her peers were content with the tradeoff of low pay with the privilege of sharing their days with the exquisite wildlife of Southwest Florida. Virtually none of my staff could afford to live anywhere near Naples, the affluent city in which they worked. Above the surface, Rookery Bay is simply gorgeous, an icon of tranquility and a flourishing font of marine and bird life. But, like much of the ocean, the beauty of its surface belies the mayhem that takes place below as a result of the dramatic hydrological changes of the lands surrounding it.

Farther south, Florida Bay lies between the Florida mainland to the north and the Florida Keys to the south. Also historically an estuary and once described as having "gin clear waters," it is now a pale, muddy green, its nurturing headwaters from the River of Grass strangled to a trickle. Shannon shocked the class—and me—by informing us that at that moment, the salinity of Florida Bay was actually higher than the ocean waters it would eventually meet.

Thankfully, as the Everglades restoration plan was developed, it was recognized that the Everglades ecosystem did indeed stretch beyond the bays and estuaries of South Florida to the coral reef tract south of the Florida Keys. Restoration considered sawfish as much as it did sawgrass. And so, one of the more unlikely—and expensive—components of Everglades restoration was born. Tens of thousands of septic tanks and illegal cesspits in the Florida Keys had been leaking through the porous limestone for decades. Nutrient-rich human waste was increasingly making its way to the reefs, fueling the growth of macroalgae. Equally serious, the leaking waste carried numerous human pathogens, which were now known to infect corals. And so, in 2000, a daunting effort was initiated to construct a massive $1 billion sewer system stretching more than 100 miles across 42 islands from Key Largo to Key West. To the corals of the only living coral barrier reef in the United States (and the third largest barrier reef in the world), the sewer system is a centerpiece of the restoration effort. It remains to be seen if the project will succeed in restoring the devastated reefs of the Keys, suffering from ever-warming waters, nutrient-fueled algae, and the continued absence of *Diadema*, and still recovering from a history of overfishing of algae-loving herbivores. But addressing one of corals' deadliest enemies—nutrient pollution—may well offer the best new hope for their restoration and survival.

CHAPTER NINE

IT'S NOT EASY BEING GREEN

Farming looks mighty easy when your plow is a pencil, and you're a thousand miles from the corn field.

—U.S. President Dwight D. Eisenhower

Not One Drop to the Sea

For years the agricultural sector in Cuba was dominated by the United States. That ended abruptly with the 1959 revolution, which saw farmland seized from most private owners and nationalized. The state owns 80 percent of the land and leases most of that to farmers and cooperatives. The remainder is owned by private farmers and cooperatives. The country has employed various models, from the Soviet-style large-scale, industrial collective farms, where farmland is aggregated and run directly by the central government, to cooperatives, where member-owners collaborate and can even share in some of the profits.

Agriculture has been a key part of the island's economy, from its felled forests for shipbuilding to its world-renowned tobacco. Trying to capitalize on its enviably rich, fertile soils, nearly 20 percent of Cuba's workforce is dedicated to agriculture. But for decades, the sector has fallen far short of expectations, now accounting for less than 10 percent of the country's gross domestic product. Remarkably, Cuba imports up to 80 percent of its food. Nearly 40 percent of its agricultural land lays idle, much of it overrun by exotic, invasive plants.

Beginning in the sixties, much of the alliance with the Soviet Union focused on agriculture. The Soviet Union supported Cuban agriculture by paying premium prices—more than five times the world market price—for Cuba's main agricultural export, sugar. As much as 95 percent of Cuba's citrus crop went to countries belonging to COMECON (the Council for Mutual Economic Assistance), comprised of most Eastern Block countries and led by the Soviet Union. In return, the Soviets provided Cuba with nearly two-thirds of its food imports and 90 percent of its petroleum.

With a strong, unwavering Soviet wind at its back, Cuba was on its way to replicating the Soviet model of agriculture, a system highly mechanized, fertilized, and centralized. Cuba used more than 1 million tons of synthetic fertilizers and up to 35,000 tons of herbicides and pesticides per year. In addition, the growing collective agricultural apparatus consumed enormous quantities of petroleum, fuel for tractors, trains, and distillates to manufacture pesticides. Expanding agriculture was a priority and a regular source of boasting by Fidel Castro during his legendary public appearances. The Mexican magazine *Sucesos* (Events) covered a 1966 public speech by Castro, conveying more an atmosphere of a religious revival than a political discourse on the state of the nation: "Women, children and a few old men were shouting at the top of their lungs: 'Fidel, Fidel!' Merriment was everywhere. One young mother with her child in her arms wept for happiness."

Cubans' relationship with Castro was complex and double-edged. On the one hand, there was discontent with the struggles of day-to-day life, the state of the economy, lack of opportunities, oppression from the regime, and inability to travel abroad. On the other hand, Castro held a caring, charismatic, and fatherly place in Cubans' hearts. They felt comforted in the warm cradle of his arms, convinced he cared deeply about the Cuban people and held an unwavering commitment to making their lives better. He was masterful at conveying such emotion to his massive audiences and shifting blame for Cuba's woes to the U.S. embargo against Cuba, but perhaps disproportionately so, critics calling it a sleight of hand to direct attention away from measurable dysfunction of the Cuban government.

Several years before Castro stepped down in 2016, still immersed in the much-chilled relations between Cuba and the United States during the George W. Bush administration, I was swept up in a massive crowd leaving Hotel Nacional in the Vedado section of Havana. The crowd swelled along Havana's Malecón—its famous seawall—as the Gulf of Mexico tossed its wave crests over the wall into the street. Tens of thousands of small Cuban flags waved in the crowd as Castro took his place on a stage in a plaza deliberately built directly in front of the U.S. Interests Section. The Cubans called it the *protestodromo*, the verbal arena of U.S.- Cuba political conflict.

On the opposite side of the road from the embassy's interests section by the Malecón stood a poster of George Bush appearing as Hitler. It also contained the disturbing iconic image of the hooded prisoner at Abu Ghraib prison, the facility on the outskirts of Baghdad where American treatment of Iraqi prisoners faced international protest.

At the far end of the plaza a statue was erected of the beloved Cuban hero José Martí. He was cradling a young boy in one arm, the other outstretched pointing at the U.S. Interests Section. (When diplomatic ties were cut between the United States and Cuba, the U.S. embassy in Cuba became the U.S. Interests Section, which represented U.S. interests and functioned as a scaled-down embassy. Similarly, Cuba's embassy in Washington, DC, became the Cuban Interests Section.) The boy in the arms of José Martí looked suspiciously like Elian González, who survived a journey by raft to Miami in the late nineties. Tragically, his mother did not survive the crossing. An enormous controversy ensued among the Cuban American population in Miami when the boy was ultimately returned to Cuba to be reunited with his father. The statue was meant to convey a warning from Martí to the boy about the imperialist Americans in the building he pointed to and what it symbolized. However, among ordinary Cubans, a popular irreverent interpretation of what Martí was saying to the boy differed slightly. In their version, Martí is explaining, "That's where the visas are." The streets that day on the plaza were lined with buses—thousands of people were bused in from surrounding towns before dawn. Attendance was mandatory. Anyone not participating was subject to arrest.

Forty years earlier, during that 1966 speech, Castro responded to the wild applause of the crowd with his signature charismatic embrace of the spellbound audience: "This spirit is what made the Cuban Revolution grand and this spirit urges us on to accelerate the pace, to work with greater enthusiasm to solve the needs of the people!" More applause. The example he then upheld to his audience was the one that preoccupied Cuban citizens and political discourse: food. "300,000 new hectares of land are being put to productive use each year, for pastures, cane, orchards, legumes, tubers, grains, cotton and other crops. And this figure will increase in the coming years," Castro boasted.

But what he said next was chilling—if you're a marine scientist. Castro proclaimed, "We also plan to develop all our water resources with the goal of not permitting one drop of water to reach the sea. . . ." A ringing applause followed, according to *Sucesos*'s account. The plan's goal was to ensure that Cuban crops would receive the irrigation they needed, in turn increasing production. But if taken to its logical conclusion as expounded by Castro, and not allowing a single drop of fresh water to reach the sea, the result would be nothing less than catastrophic. Like Florida Bay, the brackish estuarine waters would become as saline as the surrounding marine waters, a death sentence for the nursery grounds for countless species, including, ironically, many commercially important fish species, which, of course, are an important component of the food Cuba was so desperate to provide to its people. The "not one drop" plan had the stench of over-the-top grandiose thinking born from the hubris of Soviet planners. For decades, the Soviet Union was putting into motion an overambitious project to reverse the direction of Siberian rivers draining valuable water "uselessly" into the Arctic Ocean to the north while those precious waters could be redirected south to irrigate the dry agricultural regions of Soviet Central Asia. The plan ultimately collapsed in the mid-eighties and, thankfully, Siberian rivers continue to roar northward.

In the end, Cuba constructed numerous dams and reservoirs, many still present today. Fortunately, the endeavor fell far short of the goal of holding back all fresh water. Nevertheless, according to a number of colleagues, including Cuba's former head of fisheries, Dr. Julio Baisre, and Gaspar González, lasting damage was caused by the alterations of river flow that had occurred. They both cite serious drops of pink and white shrimp populations that depend on Cuba's southern estuaries. These same scientists—and I—wonder if in some perverse way, limiting river flow ultimately proved a benefit to coral reefs, especially during the fertilizer- and pesticide-rich days of Soviet-era agriculture. Sacrificing one ecosystem to save another is among the agonizing decisions we face with the urgent triage that defines 21st-century environmental issues. In the end, perhaps this was, from the perspective of a coral colony, a happy accident that allowed it to thrive these many years, unencumbered by the nutrient pollution that left so many other Caribbean reefs smothered by the massive macroalgae it fueled in what were nearly *Diadema*-free waters—and increasingly overfished and fish-free waters. But like so many other intriguing explanations of Cuba's remarkably healthy corals, we are left to speculate and crudely piece together and reconstruct possible scenarios spanning decades—there are simply no hard data.

Return of the Ox and the Machete

For one accustomed to the vast mechanized industrial monoculture farms of wheat and corn of the U.S. Midwest, a drive into the heart of Cuba's agricultural lands is jarring. Mile after mile, a landscape unfolds that is scarcely evolved from that of 19th-century farms. Ox-drawn plows, slowly struggling their way through copper-colored dirt, have replaced long-abandoned Soviet tractors seemingly melted into the moist Cuban soil. Rusty machetes, burning backs, and human sweat have replaced harvesting machinery. The Special Period's transformation of Cuban agriculture endures in the 21st century. Cuba's economy, until the pandemic of 2020–2022, had grown stronger, albeit modestly and precariously. Facing the dire economic difficulties of the Special Period, the Cuban government came to the conclusion that revenue from one of its most important exports, sugarcane, waning in production and facing falling global prices, would fall far short of keeping the fragile economy alive. Neither could its other primary crops, tobacco and citrus, fill the economic void. Reluctantly, the country turned to tourism, today the second-largest pillar of the Cuban economy. Sugar mills have been closing since the mid-nineties.

Margarita Fernandez and I have followed very different paths in Cuba. Margarita, executive director of the Cuba Agroecology Institute in Vermont, plods through the mud of Cuban farms while I glide in clear waters above Cuban coral reefs. And while it might seem that we have nothing in common, the fact is that our work intersects at one of the most important focal points affecting both sustainable agriculture and healthy coral reefs: fertilizer. It has forged a critical link between unlikely allies around nitrogen, an element that on one hand fosters life on land, on the other extinguishes it in the ocean. Thus, Margarita's work with her Cuban colleagues toward sustainable farming practices—enhancing production of crops using less fertilizer and pesticides—serves the needs of both sugarcane and angelfish. Margarita filled my inbox with fascinating scientific reports, vividly describing the agricultural situation during the era of Soviet-supported agriculture in the mid- to late eighties. At that time, she explained, Cuba had the highest level of fertilizer and pesticide use, irrigation, and tractors per hectare compared to the rest of Latin America.

A New Jersey native, Margarita has Hispanic roots from the Dominican Republic and Spain and grew up within the Caribbean diaspora community there. As a student she was fascinated with ethnobotany—the study of traditional practices of different regions relating to plants, their cultivation and use—and became increasingly interested in Cuba. She was deeply inspired by Cuba's remarkable agricultural adaptation in the post-Soviet era, and she resolved to go to the country. "I was obsessed with Cuba," the intonation of her voice revealing a personal as well as professional passion. She visited Cuba in the late nineties, lived in the country for a time, and continues her important work there to this day, critical work that is supporting Cuba in charting a sustainable, organic path for agriculture.

Cuba's organic agricultural achievements have been impressive, notwithstanding the fact that they were adopted out of necessity and desperation. Today, perhaps 70 percent of Cuba's agriculture could be considered organic. Over the years, new methods for composting, use of earthworms, integrated/biological pest management, and other methods have advanced and have received worldwide recognition as an example that crops can achieve higher productivity using fewer resources—especially fertilizers and pesticides—mitigating environmental impacts. And this is where Margarita's work has focused, helping to build capacity and new farming methods, and hosting farmer-to-farmer workshops, with successful—and sustainable—results. "It is really inspiring and really incredible to see," she says. Indeed, despite the monumental challenges, Cuba has largely succeeded in reducing hunger and malnutrition, though the average Cuban—still depending on ration cards for a modest quantity of staples that don't come close to providing enough food—would say without hesitation that there is a long way to go.

Hot Town, Farming in the City

A Cuban farmworker, glazed with sweat, parked the wheelbarrow in front of the American group I was leading. The wheelbarrow's tire—and our shoes—were muddied by Cuba's trademark rich, pungent, copper-colored soil, considered some of the most productive soil in the Caribbean. Under a swelling morning sun, the worker briefly removed his straw hat, wiping his brow with his shirtsleeve, before beginning his task of slowly removing each head of lettuce, carefully inspecting each, meticulously removing any wilted outer leaves and brushing away any dirt. He then tied them up, placed them in a basket, and deemed them ready for market. We watched under a brightening sky as the lettuce glowed with an iridescence as if illuminated from within, each head a fireball of brilliant green, beautiful and surreal. It screamed fresh, flavorful, healthful, organic. The lettuce we ate (or chose not to eat) at the hotel was, in comparison, limp, pale, and flavorless. It seemed incongruous that with such farms less than an hour from central Havana, this lettuce—along with many other delectable vegetables—would not make its way to the buffets of the hotels. Alas, most hotels were still supplied by the centralized food distribution system.

As the farmworker was joined by another, I began snapping photos. I inspected the camera's LCD screen, as if I needed to confirm that the vivid color of this lettuce I was seeing with my eyes was genuine. Sure enough, my camera captured the same astonishing green. Today I show these photographs in many of my presentations, feeling obligated to explain to my audience that the colors are genuine, and no Photoshop trickery was involved. That day we were visiting the Organopónico Vivero Alemar outside Havana. A proud Miguel Salcines, who oversees the complex, provided a fascinating tour of one of Cuba's premiere *organopónicos*, an urban farm cooperative, but uniquely Cuban and free of industrial fertilizers and pesticides. *Organipónicos* were a necessity born of the Special Period as a community response to faltering food supplies, not unlike the so-called Victory Gardens that emerged in backyards, empty lots, and rooftops in the United States during World War II when many foods were rationed. Nearly 20 million Americans participated, many forming cooperatives with their neighbors.

Organipónicos have been a great Cuban success story, another example of what Margarita finds so inspirational about Cuba's evolving model of agriculture. As cooperatives, they are publicly run and owned. Participants can share in a portion of the profits, albeit within limits imposed by the government. Born in Cuba and now replicated around the world, *organipónicos* have taken Victory Gardens to a new level: They are fully organic and tightly woven into Cuban society. A typical *organipónico* is characterized by drip irrigation, polyculture and crop rotation, composting, natural fertilizer, and biological pest control. They have emerged in vacant lots, including former parking lots; abandoned building sites; and along roadways.

Miguel knelt at the end of a row of bright red tomatoes as our group gathered around. His fingers gently stroked the stems of the colorful cluster of orange marigolds planted at the end of each row of crops. He explained how they serve as a natural form of pest control that keeps insects and nematodes away. The tomatoes and other vegetables we saw were robust, healthy, and free of pest damage.

He then led us to a small structure where he reached into an awaiting bucket and approached us with cupped hands overflowing with large earthworms, passing them in front of each of us. He smiled as he presented us with the wriggling mass—his smile broadening slightly as some members of the group recoiled at the sight. He explained the importance of earthworms, how they help to mix organic matter into the soil, improving soil structure and the infiltration of water. Beyond the marigolds, earthworms, and other practices we saw, it soon became clear a magic ingredient essential to the growth of such an exquisite bounty of vegetables was human labor. An *organipónico* is a labor-intensive endeavor, but in a country where labor is especially inexpensive, it works. Since its inception, the model has grown dramatically, cheered on and supported by the government. The urban agriculture sector produces tons of vegetables, fruits, and herbs without the use of chemical fertilizers. Cubans know that the best produce can be found at the *organipónicos* in their neighborhood. But they know it's also the most expensive produce available.

Take Me to the River

Cubans don't compete with their neighbors for the greenest lawn. Most don't even have lawns, at least not in the American or British sense. The few manicured lawns one sees in Cuba surround embassies and hotels.

Meanwhile, most of Cuba's agriculture—in its vast farmlands and in the heart of its urban centers and small towns—is organic, free from industrial fertilizers and chemicals. Nor has Cuba created an Everglades-scale transformation of its lands, replete with canals and pumping stations, drastically altering the flow of water across the land and increasing nutrient load as it enters the sea. In fact, Zapata Swamp—often referred to as the "Cuban Everglades"—lies virtually pristine along Cuba's southern coast, adjacent to the Bay of Pigs. Covering an enormous 1,681 square miles, it is the largest protected area in the Caribbean, considered the best-preserved wetland in all of the Antilles, part of the larger Ciénaga de Zapata Biosphere Reserve and designated as a "Wetland of International Importance" by the Ramsar Convention on Wetlands. The corals that lie off its shores are healthy and spectacular and close to Fidel Castro's private island refuge that Jean-Michel Cousteau spoke of and had an opportunity to visit. On a boat in the middle of Zapata Swamp, amid the quiet and the sound of distant birds, I could see only tall swamp grasses to the horizon in any direction I looked. I wondered if this is how the Everglades looked before the first steam shovel tore its back.

Given all of these factors, it stands to reason that when the summer rains sweep over the land and their waters trickle into Cuba's streams and rivers and eventually find the sea, the nutrient load is modest. The macroalgae that have slimed and smothered Florida's reefs—and those throughout the Caribbean—aren't present, at least in part, because the reefs in Cuba aren't bathed in a nutrient soup. Algae growth on corals is limited and unable to overgrow and lay waste to the coral reefs, evidenced by proud stands of elkhorn coral, perfect spheres of brain coral, and bright purple sea fans waving with the tide. The colors sparkle as light streams through the crystal waters above, the sun dancing upon the gentle waves. The corals are clean and slimeless.

Organic agriculture and minimal nutrient runoff is an explanation many of us have cited as to why Cuba's reefs thrive. It's widely accepted. It's logical. It fits. It's hopeful. It's inspirational. But is it true? We have been largely inferring and speculating based on our experience elsewhere. But in the end, there has been no hard data to point to. Until now.

"My luggage was packed with like a dozen of the biggest Reynolds Wrap rolls," geologist Dr. Paul Bierman of the University of Vermont recalls from his first trip to Cuba in 2018. Packed in with the aluminum foil was another important piece of scientific gear: Vermont maple syrup. Clearing customs was cumbersome.

His experience shows that an essential part of our work as scientists is to bring bulging suitcases full of scientific gear needed for our work and often requested by our Cuban colleagues. The challenge is that customs officials aren't scientists, and sometimes after an hour and a half of awaiting our luggage, we are hauled into an inspection room where we must gird ourselves for a lengthy explanation. On one trip, I brought four large dark brown gallon jugs of calibration fluid for a Data Sonde—a device that measures salinity, pH, turbidity, and other characteristics of seawater as it is lowered to depth in the water—which we were using on the boat during research expeditions. The calibration fluid provides known concentrations of salt and other chemicals used to calibrate the Sonde and ensure its accuracy. It wasn't until I offered to drink some of the fluid that Cuban customs agents decided to say, "*Bienvenidios a Cuba*," and release the nutty professor from his confines and allow him into the country with his jugs of salty water. I became surprisingly accustomed to these interrogations.

It is forbidden to bring GPS units into Cuba, a restriction that continues today, making little sense since the advent of smartphones, freely allowed into Cuba, all of which have integrated GPS receivers. For several years I would carry a white matchbook-sized GPS tracker that allowed me to mark each photograph I took with the precise GPS coordinates, critical for some of our investigations. I would wrap white iPhone headphone wires around the unit and place it in my carry-on luggage. Pulling everything out of my bag, an inspector would look at it, proclaim it an MP3 player, and move on to the next item. In the event of an emergency, I carry an emergency locater beacon on my trips since we spend time offshore. When it is activated, NOAA receives an emergency signal via satellite and puts into motion a rescue operation. Of course, the unit contains a GPS. I attach it to my scuba regulator with duct tape and it nicely blends in as an integrated part of the unit. I've never been questioned about it. The most upsetting experience I've ever had entering Cuba was when I was carrying several chocolate bars that, like Paul's maple syrup, were gifts for Cuban friends and colleagues. Under the guise of national security, the guards proceeded to sample the chocolate, using their mouths and taste buds, taking an ample sample to ensure it was indeed chocolate and not a nefarious substance meant to overthrow the government.

U.S. customs can be challenging, too. On occasion I would need to bring small biological samples into the United States. I recall bringing a small vial of preserved amphipods—tiny crustaceans—from Cuba to the States for analysis by Dr. James Thomas, a close Seacamp friend and professor at NOVA University. I tucked the vial into my toilet kit, and I suppose it passed for a small sample of cologne, "Eau de Crustacean." I was never questioned.

"For the past two and a half years, we and our team of U.S. scientists have been working with Cuban geoscientists to understand the environmental and water quality effects of progressive agricultural policies in Cuba," Paul writes in a 2020 article for theconversation.com. He cites a recent published study he and his team have completed, showing that "Cuban rivers are cleaner than the mighty Mississippi. Why? Because Cuban farmers practice organic farming and conservation agriculture to reduce soil erosion and nutrient loss. In sum, Cuba is doing a better job than the U.S. at keeping farming from hurting its rivers, and its results offer useful lessons." In partnership with the Centro de Estudios Ambientales de Cienfuegos, Center for Environmental Studies of Cienfuegos, the team of scientists analyzed water samples from 25 rivers in central Cuba, finding that the concentration of dissolved nitrogen was half that of the Mississippi River (0.76 milligrams per liter of river water versus the Mississippi's 1.3 milligrams). They also found much lower rates of erosion—a significant contributor of organic material, including nitrogen, to the Mississippi—and less sediment finding its way into nearshore waters, which can smother corals. "The loads of the rivers is now about a quarter of what the U.S. rivers are carrying to the Mississippi," Paul explained to me. I realized this was the first hard data available to support the cause-and-effect hypothesis of the widespread practice of organic agriculture on nutrient pollution. While Paul's work has succeeded in comparing Cuban rivers to American rivers, it falters on an important research need: How do Cuban rivers today compare to Cuban rivers of the past? He points out that, unfortunately, there's no historical Cuban data by which to make the comparison.

Fueled by a grant from the U.S. National Science Foundation, long days in the hot sun, and sweet Vermont maple syrup, Paul's research gained international attention as an attestation of how organic farming practices play a significant role in water quality, with, of course, critical implications for coastal seagrasses, mangroves, and coral reefs. At the same time, Paul is quick to point out that Cuba isn't an organic farming paradise. Paul's work was frustrated by the number of dams his team encountered. "There's so much damming. We have to move our sample sites because they're underwater." I told Paul that I suspected many such dams could be a result of Castro's "Not One Drop to the Sea" initiative, which he had not yet come across in his historical research. Their samples revealed bacteria in the water, which DNA analysis later revealed was from cows, regularly standing about in the cool water upstream. Paul also recounted the first time he toured Cuba's farmland. "There was an awakening. The first time we drove for eight hours, I realized, wait a minute . . . fertilizer bags!" They observed old farmers with bags of nitrate fertilizer, hand-spreading it on their fields. "So there there's still fertilization going on. You know, there's the internet stuff with the beautiful organic agriculture." But he reflected that the true state of agriculture is more nuanced—its future still uncertain.

The Salad Fork in the Road

As it struggles toward food security, Cuba finds itself at a crossroads, conflicted about which path to take. One path, of course, is sustainable agroecology. By largely sitting out the rampant development of industrial farming that has swept the world, Cuba finds itself with a unique opportunity to leapfrog into 21st-century agroecology and organic farming, benefiting from methods that didn't exist decades ago. The other path, of course, is a return to the push toward industrialized farming of the Soviet era. Margarita points out that, curiously, there is now a resurgence of interest in reviving Cuba's sugarcane industry, advocated by economists and planners at high levels within the Ministry of Agriculture. This faction has growing aspirations to once again be a leader in sugar production, that sugar could again be the country's "sweet spot." I'm not sure if Margarita or the planners intended the pun. Such a scenario, of course, would mean thousands more acres of sugarcane monoculture bathed in fertilizer and pesticide. Fortunately, at the same time, there is opposition from other high-level officials who feel that this isn't the right course and that diversifying the agricultural sector is a better course of action.

Three decades after the loss of Soviet support, Cuba still remains challenged to replace the steady stream of Soviet fertilizer, pesticides, petroleum, and farm equipment that flowed into the country. The occasional shoulder bags of fertilizer that Paul's team observed are evidence of the underlying incentive among Cuban farmers to use whatever means possible to improve their harvest. But most farmers have not had access to fertilizer for 30 years. Adapting to such challenging conditions, Cuban farmers have done what all Cubans have excelled at over the years: innovate and adapt. At times Cubans' ingenuity during hard times is jaw-dropping. In the case of agriculture, Cubans have simultaneously retreated to farming practices hundreds of years old while slowly adopting, implementing, and innovating modern methods in agroecology. In the end, the choice between industrial and organic farming may not be decided in five-year plans born in central-planning government conference rooms. The Cuban economy may well overtake any such planning. Indeed, the Cuban economy has played a dominant role in shaping which path Cuba has taken. Margarita points out that historically, as the economy has improved and the government can better afford fertilizers, the environment suffers. Conversely, when the economy falters, the environment thrives. It's a paradox that extends well beyond agriculture.

PART THREE
FORBIDDEN FRUIT, FORBIDDING POLITICS

There is a charm about the forbidden that makes it unspeakably desirable.

—Mark Twain

CHAPTER TEN
THE MAN WITH THE SILVER BRIEFCASE

A great deal of intelligence can be invested in ignorance when the need for illusion is deep.

—Saul Bellow

High Tea with the Queen

"Are you serious?" A nervous smile of disbelief swept over my face. I shook my head as I listened to a set of instructions given to me over the phone that seemed to be meant more for James Bond than for a run-of-the-mill marine scientist without a license to kill. Mayra Alonso, then Cuba coordinator for Marazul Charters, the largest of a handful of air charter services to Cuba from the United States, was delivering the complicated logistics I needed to follow to reach Havana from Washington, DC, via Miami and Cancún. "Once you're through customs at Cancún and have claimed your bags, keep your eyes open for a mustached man in cowboy boots carrying a large silver briefcase. He has your tickets to Havana. Give him the vouchers you'll receive from us and he'll hand you your tickets."

This was the fodder of espionage films, but alas, I had no matching silver briefcase to secretly exchange with my contact. On each trip, I would scan the airport for this mysterious individual without success. And, invariably, he would appear from nowhere, startle me from behind, and address me by name, "Señor David?" I would fumble for the vouchers and we would make the exchange. I would then do an about-face and make my way through passport control to where I just had been, to await my Aerocaribe flight to Havana.

As the crow flies, it would be less than a three-hour flight from Washington, DC, to Havana. Flying to Miami, then Cancún, then Havana would sometimes become a 16-hour odyssey through four airports and endless waiting, ultimately dropping me in Havana for another hour or two to wait for the baggage to emerge on the carousel, and another 45 minutes for the taxi ride into Havana and a long check-in line at the hotel. To add insult to injury, I was usually the only one with a laptop open during each flight, trying to get some last-minute work done. The rest of the passengers were preloading with alcohol in anticipation of a weeklong party in the tropical sun. During one of these especially trying expeditions—having awakened at 3:30 a.m. in Washington to catch a 6:00 a.m. flight to Miami and finding myself in Cancún at 10:00 p.m., now 20 hours into my journey, awaiting a delayed Aerocaribe flight to Havana, I did a rough calculation, despite my foggy head. I realized that I could have flown from DC to Heathrow, grabbed the Heathrow Express into London, made my way to Buckingham Palace, had tea with the Queen, and *returned* to DC in less time than it takes to travel to a country just 90 miles away from the United States. To be conservative, I calculated that my tea with the Queen would be brief and not include scones or biscuits.

I have made such journeys to Cuba more than 100 times over the years, many through Cancún. But I also experimented with flights through Montego Bay, Jamaica, where Air Jamaica would invariably lose my bags for days. I flew through Tampa, connecting through the Cayman Islands but found that lengthy and stressful. Once I flew through the Bahamas but had to connect to Cuba's airline, Cubana, and boarded a Soviet Yak-42 that was in worse shape than the ones I nervously flew on in Russia a decade earlier. Later I discovered the luxurious option of flying Air Canada through Toronto. From Washington National Airport, the first leg was a 90-minute flight in the opposite direction I needed to fly, but I was rewarded with clean and modern McPherson Airport, where I could stuff my face with Coffee Crisps and espresso during the wait for my connection to Cuba. The flight to Cuba was a straight shot of three hours aboard a clean and comfortable aircraft. Best of all, I could check my bags from Washington, DC, all the way to Havana. It was a more expensive option, and whatever stress I avoided during the flights, I would experience with unwanted attention from airport officials upon arrival in Havana. As usually the only American among the Canadians on the flight, I was "greeted" by Cuban security upon arrival and "politely interrogated," though not an infrequent occurrence on any flight. With tiny scraps of paper and writing in tiny letters, security officials would apologetically and politely ask me a slew of questions about why I was in Cuba, confirming that I wasn't with the media—which seemed to be their biggest fear—and a range of banal questions like whether I had family in Cuba, was I bringing any gifts, any money, electronics, where was I staying, and even, at times, whether I had any girlfriends in Cuba.

Flying to Cuba through a third country isn't mandatory if you're traveling legally. However, many Americans entering Cuba illegally choose the third country approach. Such visitors were warmly welcomed by the Cubans, who especially welcomed their dollars. Cuban passport control officers would never stamp a U.S. passport so as not to get the visitors in trouble when they cleared customs back in the States.

I have always traveled to Cuba legally, but for me, flying through a third country was more about preserving my sanity. Until American, Southwest, JetBlue, and other U.S. commercial carriers began regular service to Cuba after President Obama lifted many restrictions, the alternative was a horrid experience with charter aircraft departing from Miami International Airport. Check-in began around 3:30 a.m. deep in the bowels of the massive airport, where crushing lines of Cuban Americans with plastic-wrapped mountains of baggage would wait for hours to finally reach the counter. Meanwhile, the check-in agents would take their time and flirt with other airport workers who brought them espresso, and we would wait and wait, something I would later learn is as much a part of Cuban culture as dance.

The waiting only began at the check-in counter. We would then wait at the gate for planes that were of questionable heritage and airworthiness. Some looked like Frankenstein's monster, slapped together with leftover parts of other aircraft, unpainted, and bearing no markings. Predictably, it was all too common for the planes to have frequent mechanical breakdowns with hours-long delays. On one especially memorable trip, I was taking Senator Sheldon Whitehouse to Cuba to learn about our work aboard an hours-delayed charter flight. Relieved to be on our way at last, we were dismayed when somewhere over Key West, the plane banked into a 180-degree turn to return to Miami. The captain announced that there was a problem with the windshield. In fact, a partial decompression had occurred. As I deplaned in Miami, I peeked into the cockpit. The windshield was dripping with condensation from the cold, wet air that had rushed in. How the pilot could find the runway remains a mystery to me.

Mr. Bush, Tear Down This Wall

As former president of the Soviet Union Mikhail Gorbachev entered the hotel conference hall in Miami in 2005, my daughter, Anna, born in Moscow and 19 years old at the time, was the first to greet him. They briefly exchanged pleasantries in Russian before he moved on to mingle with the American crowd that had gathered to discuss relations between Cuba and the United States. Minutes earlier, as I entered the hotel, I had seen firsthand the fury of the Miami Cuban American population, waving banners and signs, vehemently protesting this gathering. They were decidedly older, presumably having personally experienced the revolution and the nationalization of their property, forced to leave nearly everything they owned as well as family and friends, when they fled to the States. They supported and continue to support the embargo and isolation of Cuba. Many still hold out hope of the return of their property.

In a *Washington Post* editorial by Gorbachev that ran that day, he wrote, "The lack of relations between the U.S. and Cuban governments . . . has not allowed for an understanding that could benefit the citizens of both nations." He reflected on the 1985 summit in Geneva with U.S. president Ronald Reagan and said that it "did much to increase mutual understanding between our two countries," and gave a nod to President Reagan's famous 1987 "Berlin Wall Speech" where, with dramatic flourish, Reagan said forcefully, "Mr. Gorbachev, tear down this wall!" Gorbachev ended his editorial by saying, "I urge President Bush to tear down the wall of the embargo now, in order to lay the foundation for a new relationship with Cuba."

As a scientist, I had always thought of our work as politically agnostic, operating within the burden of the regulations of the embargo, but decidedly politically neutral. But by the fifth year of working in Cuba, I realized that I and other U.S. colleagues working in Cuba were drawn somewhat unwittingly into the political discourse, our collective scientific work in marine science gaining attention as one of the few areas of successful collaboration between the two countries. As I gazed at the protesters, several locked eyes with mine and yelled directly at me. It seemed surreal—surely I wasn't their enemy, a "man of science." Or was I? It was a reminder that as long as we worked in Cuba, we would need to be vigilant, as we walked a knife edge of political neutrality.

I naively thought that because we were scientists doing science, we could just keep our heads down and avoid politics. Sadly, a pipe dream. I began to awaken to the fact that politics had already intruded on science, beyond the burdensome hurdles we were forced to navigate to work in Cuba. A most outrageous example was that of the "Bulletin of Marine Sciences," published at the University of Miami and considered one of the premier publications in marine science in the world. For years, they had instituted a policy of disallowing Cubans to publish in the journal. The free flow of information between the United States and Cuba is not unlawful under the embargo. Yet when I asked a high-level representative of the university about the policy, he informed me that their attorneys advised that they adopt such a policy because allowing Cubans to publish and fostering scientific dialogue with American scientists would constitute "an illegal export of information to Cuba." It was unadulterated bullshit, an outlandish policy that inserted politics into science, transparent in its aim to placate the local Cuban American population that opposed engagement with Cuba. Two years later, the assault on scientific collaboration would broaden as new Bush administration regulations would attempt to limit academic exchanges and research.

Traveling to Cuba and encountering mysterious individuals bearing silver briefcases is one matter. Working there is an entirely different endeavor. When asked for advice for working in Cuba, I reflect on my own naiveté when first setting foot on Cuban soil. I've learned the hard way over two decades and still invariably stumble. My best advice to my colleagues is to be prepared to be in it for the long haul. Trust takes time, as does navigating logistics. Be prepared to have patience beyond reason. Don't think you can avoid politics—they will invariably capture you in their snare, yielding confounding unpredictability and a profoundly unstable foundation upon which to build your program. Embrace a shitstorm of bureaucracy, maddening mountains of paperwork, and unfathomable regulations. And take your Dramamine for a roller-coaster ride through every human emotion known to exist. In the end, I've been lucky to find more joy than despair and have found great satisfaction in creating and carrying out interesting and important projects with exceptionally warm, caring, and welcoming Cuban colleagues. Many I consider among the closest friends I have anywhere in the world. I remember September 11, 2001, only a year after I first visited Cuba. As we evacuated our office building in downtown Washington, DC, just before the Pentagon was hit, I glanced at the screen of my laptop before I packed it away. The very first emails I received to check on my safety were from Cuba.

It doesn't matter if you're an American marine scientist, a musician, or a journalist. On your journey to the island, you carry not only your backpack, but with it the weight of more than 60 years of estranged relations and an oversized tangle of regulations that have sprouted from the U.S. economic embargo imposed on Cuba after the 1959 revolution that brought Fidel Castro to power. Anything you want to bring to Cuba—whether it's your body or what's in your backpack—requires a license from the U.S. Department of the Treasury's Office of Foreign Assets Control (OFAC) and the U.S. Department of Commerce. Licenses are granted to certain categories of travel, such as professional research, the performing arts, humanitarian missions, and journalists, among a handful of others, creating a de facto travel ban to Cuba for ordinary Americans, the only country to which Americans were restricted from traveling until a travel ban to North Korea was imposed by the Trump administration.

Robert L. Muse is a Washington, DC–based attorney considered the leading expert on legal issues between the United States and Cuba, a field not for the faint of heart. Well read, articulate, and one of the most intelligent people I've ever met, he is a writer and speaker of extraordinary eloquence and precision, every statement meticulously researched and grounded in law, not conjecture, opinion, or politics. His writing is superlative, and with a deep, confident voice reminiscent of the late actor Jack Cassidy, he speaks with authority that brings a conference hall of hundreds to dead silence, eager to digest each and every word. His guidance has been essential for dozens of NGOs—including Ocean Doctor—to establish their programs in Cuba and navigate the unfathomable, complex maze of regulatory issues they face. He is the one I trust most to explain the unexplainable—the U.S. embargo on Cuba.

Robert explained that although it has been altered over time, most of the economic embargo imposed upon Cuba by the United States is intact from its inception in 1963. The regulations "have two principal features: One, anything that a U.S. national does, that has any financial aspect to it is considered a transaction and is illegal unless it's authorized by the U.S. government. And it's subject to both criminal and civil penalties. The second aspect of it are export controls, which operate worldwide on any item of U.S. origin. It doesn't matter how many countries it passes through or how long ago it left the U.S." It's stunning and not well known that the embargo has worldwide implications. "Say a European NGO wants to donate computers and copiers to a Cuban NGO. That is unlawful . . . unless they can demonstrate [that each item contains] less than 10 percent of U.S. content, or go through the laborious process of getting an export license." Today it's almost impossible to determine the percentage of U.S. content in computers and other items assembled from parts sourced from around the world. "So much of the difficulty and ambiguities of the embargo are deliberate. They're meant to create an atmosphere of uncertainty where it just becomes too much. It's too difficult to work with." Many of Robert's clients are non-American, international corporations, a reminder of the global reach of the U.S. embargo on Cuba.

Although there are exceptions, "the embargo has never been a great respecter of constitutional rights. During the George W. Bush presidency, they put all the universities' programs out of business. The universities had a right to believe that their First Amendment right of academic freedom would be respected. It wasn't. It was attacked, deliberately and frontally by a group of extremely hardline embargo proponents in the Bush administration . . . and universities were being told that their programs are now illegal and subject to criminal penalties."

It's clear that Robert finds outrage in a by-product of the embargo: It goes so far as to infringe on Americans' rights to free speech and expression under the First Amendment, observing that our judiciary is "loud in small matters, whether someone can wear a T-shirt . . . when it comes to the First Amendment, but often silent when it comes to larger issues," the effects of the embargo being one of them, including Trump's abolishment and elimination of provisions that allowed for attendance at conferences in Cuba or allowing public music and dance performances and exhibi-tions of art. "I can't think of anything more genuinely obnoxious . . . that receives almost no attention." As of early 2022, the Biden administration had done nothing to reverse the Trump restrictions and most political triangulation suggests it is unlikely to do so.

While the licensing process is burdensome on U.S. and foreign companies and NGOs, and even infringing on the constitutional rights of Americans, its impact on Cuba—protracted over so many years—has been significant. Due largely to U.S. regulations, "there's not one foreign bank operating in Cuba," Robert explains. "That's because they've done a business calculus that violations of the financial transactions involving Cuba have resulted in billions of dollars of liability against foreign banks, including European banks. In one case, French bank BNP Paribas was fined over $9 billion. Other fines have been in the hundreds of millions and low billions." To make matters worse, any foreign transactions involv-ing U.S. banks that relate to Cuba can have their assets frozen, sometimes for decades. "Later those transactions come back to haunt the banks, as they go through strenuous audits that govern whether they're going to be authorized to continue banking in the United States. They have no option but to cooperate with the investigations." Inevitably, such investigations reveal suspect transactions. OFAC, which oversees and enforces these matters, "goes into its penalty mode and extracts huge settlements from the banks. So the banks end up paying the penalties. Is it worth risking their very ability to do business in the United States or subject themselves to huge penalties?" The banks choose to pay the penalties.

In my own experience, despite our meticulous attention to complying with all OFAC regulations, financial institutions have become wary—and wimpy—of anything smelling of Cuba. First it was Stripe, which pro-cessed credit card payments for donations and our Cuba travel program. Without warning, they terminated our relationship. Then it was the Co-lumbia Bank in Maryland. Without explanation, they closed Ocean Doc-tor's bank account. For good measure, they also closed my personal bank account, though I hadn't a single personal transaction involving Cuba, nor had I ever intended one.

The embargo—always referred to as the *bloqueo*, the blockade, by Cubans—does have significant impacts on Cuba. Robert points out that there are, of course, "inefficiencies and false starts of Cuban economic reform" that explain part of Cuba's anemic economy. But in the same breath he points out that "it's not an either/or proposition. You can recognize that the embargo does produce a serious harm to just ordinary Cubans, it has a knock-on effect throughout their economy and its dissuasion of trade and investment." There are those that have dismissed the embargo as having virtually no impact on Cuba except to strengthen Fidel Castro and his successors in their rhetoric against the United States. Robert disagrees. "The embargo is real. It harms Cuba in the world, and therefore, it harms Cuba's population. People can argue the merits of political transformation in Cuba. But it's just not true to say that the embargo does not cause hardship . . . in Cuba, as well as violating some basic U.S. rights." This fact is not lost on the Cubans. The first billboard one sees departing José Martí Airport for downtown Havana reads, "*El Bloqueo es Genecidio*" ("the embargo is genocide").

The question of political transformation that Robert mentions raises a fundamental question. What is the end game of a policy of 60+ years that has endured through 13 U.S. presidents? The answer is muddied by the fact that the embargo has gone through three iterations over the decades. "The original embargo was imposed after the nationalization of major American corporations in Cuba. American companies controlled the telephone company, the electricity company, anything of any consequence in Cuba in the industrial infrastructure. Eisenhower put the earliest embargo on one-way trade. Kennedy institutionalized it in 1963, but it was meant to compel Cuba to pay compensation for those expropriations." During those early years, the Bay of Pigs invasion, Cuba's new relationship with the Soviet Union, and the Cuban missile crisis took the embargo into the national security realm. Robert recounts that by the 1980s, "the Reagan administration conceived Cuba as a genuine foreign policy problem because of its support for groups like the Sandinistas in Nicaragua." (Cuba's support of the communists during the war in Angola was similarly in direct conflict with U.S. policy.) "That's when the principle of the embargo became purely national security."

During the 1990s, Cuba's national security threat waned. Robert explains that it was then that the embargo "began to morph into a political transformation mechanism." Translation: regime change. "If you can drive the Cuban economy to a point of despair, the whole country explodes, like a pressure cooker, everything hits the ceiling. And how do we do it? We ratchet this embargo down so tight, the misery is so great that the people revolt. And then in the mysterious ways, this is always supposed to happen with our foreign ventures, a liberal democracy flowers and goes forth in the world. And this is where we remain today." In response to Cuban fighter jets shooting down two private planes operated by a Miami group, Brothers to the Rescue, that entered Cuban airspace, ignoring warnings from Havana air traffic control, Congress reacted by passing the Helms-Burton Act in 1996, which strengthened the U.S. embargo against Cuba and extended the embargo's reach to foreign companies trading with Cuba. The Act also attempted to enshrine the embargo in law so that only Congress could lift it. (Robert's research demonstrates, in fact, that (legally speaking) the effort failed and the president actually has the constitutional authority to unilaterally lift the embargo.)

The Obama administration began to normalize diplomatic relations with Cuba, but the embargo was never lifted. Through executive orders, Obama was able to punch holes in it, and thousands of Americans traveled to Cuba. This was short-lived, however. The Trump administration reversed Obama's reforms (though diplomatic relations were not severed) and went one step further—a step with serious global ramifications—by activating Title III of the Helms-Burton Act, which had lain dormant since the bill was passed.

Title III allows a Cuban or his/her heirs to sue any person or company for the use of property they had once owned that was expropriated after the Cuban Revolution. Robert points to two especially extreme examples for which lawsuits have been lodged. If a ship uses a dock expropriated from a Cuban citizen, even for an hour to off-load cargo, they can be sued. Carnival Cruises is facing such a lawsuit for use of a dock in Havana. In another case being litigated under Title III, a Chinese company shipped wind turbines to Cuba and off-loaded them at a dock that had once been owned by a U.S. sugar company before it was nationalized. Carnival and the Chinese shipping company are said to have "trafficked" illegally in the expropriated property. The impact of that term is not lost on Robert. "It is evocative of drug trafficking, sex trafficking, and everything else that chosen verb means. It is meant to be highly derogatory to the point of inflammatory. That set a chilling effect on foreign investment. As we speak, Title III's activation is just over two years old and has produced several thousand litigations." The Biden administration has shown no inclination of returning Title III to its dormant state.

The embargo, though showing every sign of a long-failed policy, has become a plaything of politics. Regime change and the embargo are strongly supported by the Cuban American population in the Miami region, and lifting sanctions threatens that important vote in presidential elections, a factor that clearly has figured into political calculus of administration after administration. But in the end, Robert and many others find it highly unlikely that Cuba's Communist Party will collapse under the weight of the embargo. It's demonstrated virtually no cracks for more than six decades and has endured extraordinarily difficult economic circumstances—including the Special Period of the 1990s and the pandemic that has decimated Cuba's economy in 2020–2022.

In the end, for those of us working in Cuba, the embargo is an ongoing impediment to our work, made more grave by the Trump administration's additional restrictions, perpetuated in the Biden administration. We are restricted from staying in virtually any Cuban hotel and not permitted to conduct professional meetings in Cuba. Working around such restrictions presents enormous obstacles. Robert emphasizes the boomerang effect of the embargo—it has serious impacts on science and the arts, and from Robert's perspective, it infringes upon basic American liberties. "When a president has an instrument sitting around like a loaded gun, the Cuban Assets Control Regulations, he can tell an artist he can't exhibit his work. He can tell a musician he can't perform in Cuba. I think that it's time to end this."

CHAPTER ELEVEN

THE WORK OF A GIANT

Like music and art, love of nature is a common language that can transcend political or social boundaries.

—Jimmy Carter

Spanruslish

Of course, I gave Ana María my blessing to order beef, as I did to the others at the table. It became clear that in 2000 the memory of the Special Period remained fresh—a decade of hunger we rapidly fattening Americans might never understand. I myself was terribly ignorant that this chapter of Cuban history had even existed and felt terribly insensitive for not understanding how Cubans were still nursing their wounds when I first arrived there. Over the years, I have learned that few Americans are aware that such an inconceivable crisis of suffering for Cubans had taken place just under 100 miles south of the well-stocked pantries of the United States; breaking bread with their American friends was a special and appreciated pleasure.

Along with Ana María, I met CIM's director, Dr. María Elena Ibarra, during that first visit to Cuba. My university Spanish had rusted to the point of being practically unusable, supplanted by a spattering of Russian from my days working in the Soviet Union and, of course, the fact that I had married a Soviet. Many Cubans spoke English, though María Elena and Ana María spoke little. Their generation learned Russian—today

Cuban students are taught English. So we soldiered on in "Spanruslish," an odd combination of the three languages. We not only understood one another but enjoyed the beginning of a warm bond that would endure for many years. We all shared a deep commitment to studying and protecting the sea and this was something that would transcend the formidable politics that would constantly threaten to thwart our efforts.

There is a long and rich history of scientific collaboration between Cuba and the United States that stretches back as far as the mid-1800s. That history of collaboration has included the marine sciences as well, including the *Tomás Barrera* Expedition in 1914 focused on marine and terrestrial mollusks, and a 1914 expedition by Paul Bartsch and John B. Henderson to collect fish species. There were brief exchanges of the Cuban Academy of Sciences with the Smithsonian Institution, the New York Botanical Gardens, the American Museum of Natural History, Harvard University, and several NGOs throughout the 1980s and 1990s. But in retrospect, it seems that long-term, sustained scientific collaboration—certainly in marine science—began in the late 1990s and early 2000s and blossomed over the years following. (When speaking of scientific collaboration, Sergio Pastrana never misses a moment to boast that the Cuban Academy of Sciences was founded before the U.S. Academy of Sciences.)

After that first meeting, it became immediately clear to me that to understand CIM, and to understand marine science in Cuba, you needed to understand Cuba's Mother Ocean, Dr. María Elena Ibarra.

The White Hurricane

It was called the "White Hurricane," the "Storm of the Century." The freak 1993 winter storm paralyzed each city it passed, from the Gulf Coast to New England. But before the first snowflake ever fell in the United States, the storm was already well known by Cubans. The monstrous waves of "La Tormenta del Siglo" assailed Cuba's north-facing shoreline, destroying CIM, then located on Havana's northern shore. It was a devastating loss for marine conservation, because every marine scientist in Cuba is trained at the center. Fortunately, Dr. Ibarra, CIM's director since 1981, was not about to let a little thing like the total destruction of her center stop her from her life's mission to train the next generation of marine scientists and advance the conservation of Cuba's environment.

"Maybe for you in the U.S. it would be nothing to rebuild a building, but here in Cuba, it is the work of a giant!" exclaimed Gaspar González. He was a student of Ibarra years earlier. The task was indeed daunting, occurring during Cuba's Special Period, when economic conditions made the very idea of rebuilding the center seem absurd. But thanks to Dr. Ibarra's trademark persistence, the center was rebuilt just a few blocks away. Together with its wellspring of students, CIM stands as a monument to a woman with a vision who simply wouldn't take no for an answer. "She built an institution that has endured the good and the bad; it is well known, with respect from all over the country," said Dr. Rogelio Díaz Fernández in the early 2000s, then a CIM biologist, and for several years chief biologist for its Guanahacabibes sea turtle project, Cuba's largest and longest-enduring comprehensive sea turtle monitoring and conservation program. I joined the group to visit the project, active during still, stifling summer nights when the turtles were nesting. Draped from head to toe with clothing to defend themselves from the hordes of mosquitoes and tiny biting black insects we know as no-see-ums and the Cubans refer to as *jejenes*, I've never been so uncomfortable—and thirsty—and these students did it every night. Even knowing that, there was a long waiting list of students wanting to participate. We brought supplies for the locals and in return, they fed us. During our first night, the locals shot a wild boar, and we munched on chunks of the boiled, hairy beast for dinner. Before this project, there was no intensive monitoring of sea turtles on the main island of Cuba. Dr. Ibarra built strong ties with local schools and residents, involving them in the project, and with their help, drastically reduced poaching of turtles and their eggs.

CIM teems with approximately 40 postgraduate students and more than 200 undergraduate students. Thanks to Dr. Ibarra's leadership, the profile of marine science in Cuba was elevated dramatically during her tenure of 28 years until her death in 2009. The center maintains strong ties not only with other Cuban institutions, but has built strong international ties with universities and nonprofits abroad. With Dr. Ibarra as its matriarch, CIM felt, and still feels, much more like a family than a university. Its faculty long consisted of many of her former students, and the next generation of students appreciated Dr. Ibarra's passion and selfless dedication to her students above all else. She broke bread at the same table

Cuba's Viñales Valley at dawn, featuring its iconic "mogotes," unique, isolated formations of limestone, marble, or dolomite with steep, near-vertical sides and flat top.

A snorkeler photographs a large stand of elkhorn coral in Gardens of the Queen. Once abundant, elkhorn coral is now classified as "Critically Endangered" on the IUCN Red List. It is also listed on the U.S. Endangered Species List. In a barrier reef along southern Cuba, elkhorn coral thrives.

A large stand of elkhorn coral off southern Cuba.

A large school of horse-eye jacks slowly winds its way through the corals off southern Cuba.

A Caribbean reef shark at Gardens of the Queen. This species is classified as "Endangered" on the IUCN Red List due to overfishing. Nearly 90 percent of the world's sharks have been decimated. Gardens of the Queen maintains a healthy population of this species.

Bluestriped grunt, goat fish, and snapper pack a stand of elkhorn coral at Gardens of the Queen. Corals serve as a refuge for fish, where they are protected from predators.

A stand of stony coral dominated by mountainous star coral (*Orbicella*), one of the principal reef-building corals in the Caribbean. This stand is in roughly 60 feet of water at Gardens of the Queen.

A stand of black coral at 110 feet in the Punta Francés marine reserve on Cuba's Isle of Youth. Black corals are rarely black; they can be found in reds, greens, yellows, and other colors. They are soft corals, usually found below 100 feet. Some have been recorded nearly five miles below the surface.

A diver photographs a "small" Atlantic Goliath grouper in Gardens of the Queen. Imposing and highly territorial, they can grow to nearly 700 pounds. Until 2018, they were listed as "Critically Endangered" on the IUCN Red List due to overfishing, but thanks to conservation efforts, they have been upgraded to a status of "Vulnerable."

A large black grouper swims above the coral in the Gardens of the Queen. Black grouper are classified as "Near Threatened" on the IUCN Red List due to overfishing, as populations have been decreasing throughout the Caribbean. In Gardens of the Queen, they are plentiful and large.

The purple sea fan, a striking soft gorgonian coral, resembles a fan with delicate latticework of calcite. They are oriented so that the flat end of the fan faces the current. The fan gently waves as tiny polyps pluck plankton from the water.

A white-spotted jellyfish drifts with the currents in the warm waters of southern Cuba.

A stand of healthy brain coral in shallow waters off southern Cuba. This stand is disease-free; Cuba has been spared of many diseases that affect this species elsewhere in the Caribbean.

A Caribbean reef shark at Gardens of the Queen.

A diver shines his flashlight on a stand of black coral at 90 feet at Gardens of the Queen.

Closeup of brain coral in shallow waters off southern Cuba.

Pink vase sponges stand among soft corals off Cuba's southern coast. Like corals, sponges are colonial animals and feed on plankton. However, instead of tentacles, sponges use a pump to filter water through their cells to trap plankton.

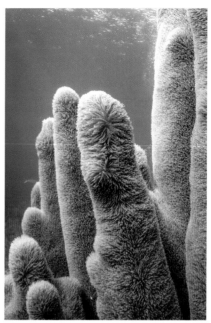

The giant barrel sponge, like this one off southern Cuba, is the largest species of sponge in the Caribbean and can reach a diameter of six feet. It has a life span of hundreds to a thousand or more years.

Among the most dramatic corals is the rare pillar coral, which extends toward the surface in parallel pillars, each draped with unusually long tentacles that sway with the current, plucking plankton from the water. The species is classified as "Vulnerable" on the IUCN Red List.

A rookery of young frigate birds nested atop a large stand of red mangroves in Cuba's Gulf of Santa Ana off its southern coast. When grown, these birds can have a wingspan of up to seven feet. Frigates soar like gliders and can remain aloft for months at altitudes up to 12,000 feet.

A trunkfish approaches a moon jellyfish above a stand of elkhorn coral in the Punta Francés marine reserve on Cuba's Isle of Youth.

The trunkfish nibbling on his prey.

A cannon stands guard at the entrance to Havana Bay with the city of Havana visible across the bay. The cannon is part of the heavily fortified Morro Castle complex built in 1589 to guard the bay.

I swim toward divemaster and award-winning underwater photographer Noel López at Gardens of the Queen. *Photo by Noel López*

as her students, traveled on the same uncomfortable bus, and slept in the same room when in the field. She refused coffee if there was not enough for her students. Despite my protests, she joined Coqui and I in pushing Coqui's Lada when it broke down. Her students loved her, admired her, and still draw great inspiration from her.

Daylin Muñoz Nuñez was a student of Ibarra's who graduated in 2001. She went on to study at Duke University and work with Environmental Defense Fund and World Wildlife Fund and is now at a major U.S. foundation in Washington, DC, where she resides. She considered Ibarra her role model when speaking of her while still a graduate student. "She pays attention to everybody. You don't have to be a doctor or an experienced person. She pays attention to young people, too." Fellow student Julia Azanza Ricardo was equally inspired. "She's a woman of great ideas with a lot of *energia*. When she has an idea she always has a way to accomplish it."

In the early 2000s when we worked closely together, Dr. Ibarra's over-packed calendar belied that, at 70, she was 15 years past the customary retirement age for women in Cuba. Among her myriad accomplishments, she helped found the Natural Botanical Garden of Cuba, is professor emeritus at the University of Havana, and was president of the Cuban Zoological Society for 16 years. She cofounded Pronaturaleza, the Cuban Society for the Protection of the Environment and from 2000 to 2009 was its president. "Prior to its establishment in 1993, Cuba had no organization whose main goal was to address environmental issues in Cuba," said Dr. Ibarra. Under her leadership, the organization was a major conservation force in Cuba.

Dr. Ibarra attributed her inspiration for education and biology to her parents. Her father ran two private schools in Santiago de Cuba, and her mother was a professor of natural science. She moved to Havana in 1950 to pursue her studies at the University of Havana, but the strikes against the Batista regime left the universities closed, so she returned to Santiago to teach natural science in her father's schools. After Fidel Castro took power in 1959, she returned to Havana and worked at the Cuban Institute for Petroleum. Following the revolution, there was a severe shortage of teachers, and the government sought volunteers to teach. She jumped at the opportunity, and taught nights at Havana schools. In 1964, she joined the faculty of the University of Havana, where she eventually became dean of the Faculty of Biology, a position she held until taking the reins at CIM in 1981.

The years have taught her that it takes more than biology to achieve conservation, and she was intent on seeing that her students' training reflected this. When we spoke years ago, she observed, "Economic, social, cultural, and political issues are all factors. Nowadays, this information is entangled for any university graduate." She welcomed students from other disciplines, hoping to build environmental awareness in the undergraduate students of engineering, architecture, and economics who work shoulder-to-shoulder with her biology students on the Guanahacabibes sea turtle project. That was by design. Her vision is truly farsighted. "Environmental education is about challenging minds, something that is very difficult if you don't plant a seed early in peoples' lives. That's why we started the Guanahacabibes project. We are there now, but what will happen there is ultimately their responsibility."

In Cuba, Dr. Ibarra faced a daily fight to keep the lights on—literally. "She barely finishes one battle and another begins," sighed Rogelio during those days. CIM had just lost its internet and email after the monthly cost was raised from $55 to $800 a month, a consequence of the bizarre mix of socialism and capitalism that Rogelio termed "Capisolism." Dr. Ibarra adds that the U.S. economic embargo has also been a serious factor.

She remains admired internationally as a leader, a director, and a visionary. But she always considered herself a teacher first, something her legion of loyal students feel intensely. "Sometimes I call her '*doctora*,' but I prefer '*profe*,' [professor]. I will always be able to learn from her," said Daylin while still a graduate student. "For me she is an example . . . I would like to be like her some day. She is *persistente*. I think she'll never give up."

I had the pleasure of knowing and working with Dr. Ibarra for nine years. She defined selfless dedication, never accepting any favoritism or special privilege, sometimes to the frustration of her students. After all, she had formerly been married to José M. "Chomi" Miyar Barrueco, the former minister of Cuba's Ministry of Science, Technology and the Environment.

She always called me "doctor" and I called her "*doctora*." But while our terms of address were formal, we also developed a friendship, and I felt privileged to experience her wonderful sense of humor as we grew closer.

Once, on a long bus ride returning to Havana from a conference, I announced that I would be buying lunch for everyone but that my organization was not permitted to purchase alcohol. Dr. Ibarra immediately stood up and loudly protested, speaking English: "Doctor! I WANT RUM!"

No Es Facil

It's a trademark expression in Cuba, close to a national slogan and certainly an understatement: "*No es facil.* [It isn't easy.]" Cubans were delighted—if not a bit incredulous—to hear President Obama say those three words in a speech during his historic visit to Havana. It's an observation, a resignation to the fact that life in Cuba is difficult, and it's impossible to work and connect with Cuban colleagues without some understanding of this. Life is difficult because of the economy, because of endless bureaucratic hurdles and the constant need to get "permission" for just about anything. It's difficult for many other reasons, including isolation from the rest of the world and the fact that countless families have been long separated since the revolution, as many have fled to the United States.

Despite their hardships, the Cubans I have known are a gentle, positive, and funny people, whose warm hearts have welcomed me again and again. I have been struck by how people help one another. It is perhaps out of necessity, but it seems more than that. There is a kindness, a peacefulness, an unguardedness, and sincerity to this culture that seems long lost in the States. I recall one afternoon in Old Havana, I was struggling to make a call on a pay phone outside a bakery. Fumbling with a phone card, I was about to give up when a passerby stopped, and without saying a word, snatched the card from my hand, punched in a string of numbers, handed the card back to me, and walked away before my stunned lips could issue a "*muchas gracias.*" The call went through.

Cubans are extraordinarily innovative and creative, and fuel the economy by working around it. "Nothing is impossible in Cuba!" a Cuban friend told me. "If it was suddenly decreed that these . . ." he looked around the room and pointed to two blouses on the bed, "blouses were illegal to have in your home, they would suddenly start appearing everywhere in people's homes." Today Cuba's black market has gone online. Revolico is essentially Cuba's Craigslist, connecting buyers and sellers more efficiently—though often illegally.

Perhaps one of Cubans' greatest assets is their uproarious sense of humor, not afraid to poke fun of themselves and their situation. On the bus to Guanahacabibes heading to the site of the sea turtle project, Julia Azanza, with her omnipresent smile and laugh, was telling a joke about someone who died and went to Cuban hell. His friends were worried about him and were able to contact him. He reassured them, "Oh, don't worry about me! There are all sorts of shortages and there's no fuel for the fires, so it's actually quite comfortable."

The New Cubans

Sadly, talented scientists, including marine scientists, have fled Cuba, many early in their careers, and we have seen a number of our close colleagues depart Cuba over the years. Clarita had a unique overseas opportunity to study at a marine biological station. Connecting through Toronto on her return, she and a Cuban colleague left the airport and headed for Niagara Falls. From there they hailed a taxi and crossed the Rainbow Bridge. Before the U.S. Customs Official on the other side could say a word, to the surprise of the official and the cabdriver, they said, "We would like to seek political asylum in the United States." Dumbfounded and unaccustomed to Cubans crossing into the United States from Canada, the official replied, "Uh, and you want to do that right now?" I believe the taxi driver was tipped well. Clarita spent several weeks in a county jail with prostitutes and petty thieves before her release. She went on to a successful career in marine conservation, as did her colleague who spent more than a month behind bars following his entry to the United States.

A CIM graduate student was preparing to attend a workshop in Belize on manatee conservation. At the last minute, he was denied permission to go, but that evening he looked at his paperwork and realized he had everything he needed—so he went. From there he began a long journey north by bus to the U.S. border. In Mexico, the bus was pulled over and officials reviewed all of the passengers' documents. Seeing his Cuban passport, they began an interrogation, demanding to know why he was there. He put his hand in his pocket and produced two manatee kidney stones. He told the *federales* that he was headed to Florida for a very important manatee workshop and that his attendance was critical. It was, of course, a fiction, but the bizarre sight of manatee kidney stones left the authorities speechless. He reached the border and asked for asylum, which he received.

Dr. Jorge Angulo had big shoes to fill. He took the reins of CIM after the passing of the beloved Dr. María Elena Ibarra. The first time I met Jorge, I could tell he had an allergy to bureaucracy. Young and energetic, he is driven and strives to get things done. He's creative, determined, and deeply dedicated. He is a pragmatist but always pushes the envelope to make things work. Over the years, we became good friends, and I have always appreciated his honesty and directness. His wife, Anmari, is one of Cuba's experts on manatees. She had an unprecedented opportunity to become the first Cuban national to attend the University of Florida as a graduate student. Jorge and Anmari moved to the United States, where they are now raising a family. Jorge is a professor at Eckerd College in St. Petersburg, Florida. But Jorge has found a way to have the best of both worlds. "I want to help Cuba," he told me. He remains very involved in Cuba, and he and Anmari continue to help advance research and education in Cuba while forging new links between the two countries. It's a rare outcome for many of the scientists who have left Cuba, but Jorge and Anmari don't feel like they've really left. They just have a slightly longer commute to CIM.

But others wonder about the situation for young professionals in Cuba today. Clarita ponders the situation, years after she left. "I honestly don't know. I mean, I think things have changed a lot in Cuba to the point that young people probably don't even aspire to go to university or be professionals." She left to advance in her professional career, something she felt she couldn't do in Cuba, let alone survive on the state salary. "We are in a different generation, far removed from whatever better situation the revolution created for the Cuban people. We already grew up in the system with free education, but didn't have enough teachers. We grew up in a system with free health care, but the health-care system is broken— there is not enough medicine. Doctors are charging on the side for the appointments and selling the medicines in the black market. So what is there for us and this new generation? The Cuban government is not offering anything else that is good for us. Why do we want to support this system that is not giving us any opportunity? So it's hard to predict what is going to happen next. I think the internet has opened up people's mind and giving them an idea of what the future could look like."

Young Cubans are eager to work and the government is slowly permitting more and more individuals to become *cuentapropistas*, private entrepreneurs. But your trade must be on the approved list. Among some, entrenched in the old system, there is a negative attitude about anyone not associated with an "official" institution. We contracted with Lis Nuñez, a Cuban *cuentapropista*, to serve as a consultant in Havana. At one point, the head of a major Cuban NGO scolded me, questioning my decision to contract with someone "*de la calle* (from the street)," the term used pejoratively. The notion of working independently is a difficult transition for Cuba, and difficult for those entrenched in the old system to accept, but it's increasingly seen by the government as necessary, an important part of the solution to help revive the Cuban economy.

Bittersweet, the Forbidden Fruit

As joyous as our work together can be, there are unexpected potholes along the road—some comical, some frustrating, some tragic. The forbidden fruit is bittersweet. The lesson of Dr. Ibarra—persistence—is key. So is patience. But many obstacles, large and small, conspire to turn one's hair white.

The realities of being a scientist or student at CIM are formidable. As mentioned, the average monthly salary for a marine scientist is roughly $20. When our collaboration started in 2000 and for years following, CIM had only one phone line. A job requirement for a CIM receptionist is to possess a very loud voice and not be afraid to use it. A call would come in and the receptionist would march to the bottom of four flights of steps and shriek upstairs, "COQUI! *TELEFONO!*" (Coqui would then need to scramble downstairs to take the call.) Internet was virtually nonexistent, available only at the main university building, painfully slow and limited. With scarce fuel for their boat and no vehicle for transportation, students walked a mile to the water with their scuba tanks on their backs, carrying what little scientific equipment CIM could afford. The practice continues today. At some point, CIM acquired a German panel truck and used it to transport students, piled into its windowless storage area. The truck was a bright orange upon which someone had spray-painted, "*Viva la Ilusion*" on its side. It was uncomfortable, but it got the students to the water. It is impossible not to admire the dedication of staff and students, working with so little to achieve so much. The quality of their work maintains rigorous scientific standards despite the hardships.

Beyond the regulatory hurdles, working together presented all manner of logistical challenges. Cuba's internet was so limited that we could not send attachments larger than 64K, a fraction of a megabyte. Word documents had to be sliced apart in order to be small enough to clear the Cuban server, and emails could take days or weeks to receive a reply depending on when the recipient could get to the university to use a computer. There was no regular mail service to Cuba, and courier services were prohibitively expensive. So were phone calls, costing between $1 and $2 per minute, even to this day. Expensive and difficult, communicating by phone or email was technologically possible, but in other ways impossible. Cubans were reluctant to write or say anything specific lest their communications be monitored. The only way to truly work together was face-to-face. Before the short-lived reforms by the Obama administration, it was nearly impossible to obtain visas for Cuban nationals, not to mention the restrictions placed on Cuban travelers by the Cuban government. The solution was for us to travel there, and so we did, which is why my passport has been stamped more than 100 times in Cuba. Thankfully, WhatsApp, Telegram, and other encrypted communications apps have dramatically improved communications, and in general, Cubans feel freer today to express themselves. However, internet communications, especially social media, remain at the whim of the government, which can shut them down in an instant as they did during the protests during the summer of 2021.

Sometimes it matters not whether you have your official paperwork in hand—you may easily find yourself at the mercy of a low-level guard, employee, or even police officer, disinterested in anything except the small world they control. For years, Cubans were not allowed in hotel rooms—they were reserved for foreigners. "Cubans discriminate against Cubans," said Clarita. For a trip to Guanahacabibes, for a sea turtle workshop we were hosting, we had to obtain special permission in order for the Cubans to stay at the same local hotel as the Americans. The desk clerk, even when confronted with the stamped, official document permitting Cubans to stay, refused. "It is okay for Americans, but Cubans, no." Finally, I stated that if the Cubans don't stay, the Americans don't stay. Fear crept across his face. He finally succumbed. But we weren't always so lucky.

After months of preparation, we had organized a major environmental economics workshop focused on Cuba's Jardines de la Reina, Gardens of the Queen, Cuba's large protected area 50 miles offshore of Cuba's south-central coast. With professors from California and Colorado along with other experts from the United States, hosted by environmental experts and economists from across Cuba, all preparations were made and permissions secured. Tamara Figueredo Martín, an environmental economist at Centro de Investigaciones Ecosistemas, Cuba's Center for Coastal Ecosystem Research (CIEC), was leading the workshop with me, and the evening before departure, we spoke excitedly by phone about the long-planned workshop. The next morning, my cell phone rang. Tamara was sobbing. Despite the fact that all of our paperwork was in order, the Cuban Coast Guard official on duty that day refused to allow the Cubans to join the Americans on the boat. He offered no reason. One of the divemasters tried to console us. "The guy wakes up in the morning. He has two shirts. One is a 'yes' shirt, the other a 'no' shirt. Today he had his 'no' shirt on. He'll say 'no' to everything."

Such encounters arise with regularity, but usually without such grave consequences. Some are comical and absurd. Coqui and I were driving through Havana when a motorcycle policeman pulled us over. Coqui told me not to utter a word. She could get into trouble if suspected to be illegally serving as a taxi for a foreigner. After studying her documents, we learned of our offense: "You are both enjoying your conversation too much. I suggest you go home, finish your conversation, and then drive." To me, this was an absolute absurdity. To Coqui, just another afternoon in Havana.

An important part of our work was connecting our funders directly with our Cuban colleagues. I always felt it was important that our supporters had an opportunity to connect with the Cubans, not only to better understand them, their work, and our projects, but also to better understand their lives and the challenges they face doing research in Cuba. Equally important, I have always felt that it was imperative that our Cuban partners well understood how foundations work, how we, the grantees, have to write our proposals, develop our budgets, and deliver what we promise and do so on time. A California NGO we brought to Cuba offered to make a very generous donation of $50,000, an enormous sum for Cuba, to support the Antonio Nuñez Jimenez Foundation. I suggested a new environmental learning center that could serve many students, host conferences, etc. Everyone loved the idea. All that was needed was a formal proposal before the California group's next board meeting a few months away. I worked furiously with the Antonio Nuñez Jimenez Foundation staff, taking photos, roughing out possible floor plans, developing concepts for the new center. We had put together much of the information, but as the days went by, I was unable to get a draft proposal from them. They asked if it would be possible to have a few more months. I pleaded, "No. This is not how U.S. foundations work. This is our only opportunity. There is a hard deadline." To save time, I desperately wrote the proposal myself. But with a month left on the clock, there was one piece missing and I couldn't provide it: We still needed a blueprint to show—at least roughly—what the building would look like. I stressed that Ocean Doctor would pay the costs. Time passed, and despite my pleas, no blueprint was ever produced. The deadline passed. The funding was forever lost. I recount this story not to lay blame on my Cuban colleagues or the Americans. The important lesson is that despite so many goals we have in common, so many opportunities we have to work together, and how well we *do* work together when things go right, we are still the products of very different cultures and very different ways of getting things done. Even after two decades, it is still a gap that I find challenging to bridge.

No matter how many times I've been to Cuba, no matter how many high-level Cuban officials, ambassadors, or comandantes I have known well, no matter what my accomplishments, dedication, and sacrifice, I am an American, and as such I live on a political tightrope, fearful of saying something that might raise the political ire of the Cuban government. I thought my worst fears were realized when I met with Liliana Nuñez, who heads the Antonio Nuñez Jimenez Foundation. We had been collaborating to launch the first environmental film festival in Cuba, a project we were both very excited about.

When I arrived at her office, I was puzzled when I saw pages of the Ocean Doctor website printed out and strewn about. She was fuming and held up the latest Ocean Doctor newsletter, which discussed the film festival. I was shocked to learn that she had taken great offense to the wording I used in our latest newsletter about the role of film in helping to strengthen environmental awareness for the Cuban people. I would have used the same language for a film festival meant for Americans. Liliana's interpretation of my well-meaning words was dispiriting. To her, implicit in my statement was that Cubans lacked environmental awareness. Further, she saw it as an attack on her organization, that they had failed to provide environmental education to the people of Cuba, one of their major activities. She yelled for an hour as her colleague continued to print pages of our website.

As she issued the words, I could scarcely believe I was hearing, the pit of my stomach fell, leaving a void of darkness and despair I had never felt before: "This is COUNTER-REVOLUTIONARY!" she shrieked. My spirits sank. The 16 years of trust I had built was irrelevant. I hadn't meant any offense and certainly didn't intend the meaning she extracted from my words, but there was simply no way to present my side of the story. I went as far as to concede that I might have chosen different words, that we had a misunderstanding from which we could move forward. Her eyes narrowed and her face reddened as she pointed her forefinger at me and said, "No!" Her voice raised. "It is YOUR fault!" I left the building observing the pitying eyes of my colleagues at the Foundation who had obviously heard every word. Panicked that this would spell the end of my work in Cuba, I hastily made an appointment with Josefina Vidal, the head of U.S. relations at the Ministry of Foreign Relations, and one of the architects of the normalization of diplomatic relations between our countries. She quietly reassured me that all would be okay. When Liliana and I next saw each other at a meeting in New York, we spoke politely and began the process of rebuilding our relationship.

Though I've experienced many challenges on the Cuban side, the Cubans have their own anecdotes about their challenges with Americans. Sergio Pastrana recalled he had once been invited to the United States as a representative of the Cuban Academy of Sciences. Upon arrival in Miami, he was immediately detained. " 'Why are you here?' they asked. I said, 'I was invited to be part of a meeting.' They said, 'That's easy to say. . . .' I said, 'No, it's the truth!' They locked me in the room and went to lunch for two hours. If only I had been thrown in jail, I would have become a national hero!" He then recounted another story about an incident in 1995. The vice minister of the Ministry of Science, Technology and the Environment (CITMA), was invited to the United States for a conference on El Niño. "The invitation came on White House stationery. It made a big buzz in Havana. He went, but was thrown in a jail. They took his belt and shoelaces. The vice minister was thrown into a cage!" Sergio exclaimed. "After frantic diplomatic efforts all night long, including knocking on the door of someone's home in the middle of the night, he was released at daybreak and he attended the conference."

And then there are incidents that are a cold slap across the face, matters of life and death that put things in perspective. His name is Eduardo Alonso Ramos, but everyone calls him "Alonso." He was supposed to join me for a final meeting at Havana's Marina Hemingway in final preparations for an expedition that he was to be part of. He never arrived. That's certainly not unusual in Cuba. Transportation is always a challenge. Perhaps Alonso's motorcycle broke down. Perhaps he couldn't get gas. Who knows? I wasn't worried. Though it was a bit unusual that he never called, even that night. The next morning, I learned the tragic truth. Alonso was filling a scuba tank at CIM when it exploded, killing him instantly. He was only 41, survived by his wife of 36. They buried him that morning. Like the rest of us, Alonso absolutely loved the sea. It was his life. When we met, his eyes were wide with excitement about getting out on the water with us. He was a sailor, a divemaster, an expert technician. Word of his death had already spread through the marina by morning.

I went to CIM to pay my respects to my colleagues. I could hear sobbing inside the building from the sidewalk. Seeing me arrive, they cried some more. We embraced. Representatives from the Ministry of Interior and the police were already there, examining the accident scene. There was no time to grieve. It turned out that the tank didn't belong to CIM. He filled it as a favor for a friend. The 35-year-old tank had never been hydrostatically tested for metal fatigue. Jorge Angulo, who was CIM director at that time, faced a yearlong investigation that could have seen him behind bars. He was a different person that year, visibly worn down by unrelenting stress. Thankfully, Jorge was exonerated from any wrongdoing.

The following day, before departing for the States, I made my way down to CIM's basement, where the accident had occurred. I looked at the rack of CIM's scuba tanks there. The tanks looked ancient. With the trauma of Alonso's death so fresh, I was haunted by the thought that this could easily happen again, so much so that I asked Ann Luskey, a champion of marine conservation and the owner of the boat that we were to use for the expedition, if I could donate the 11 brand-new scuba tanks stowed aboard to CIM to replace their old ones. She graciously agreed and we off-loaded the tanks. In doing so, I was breaking the law—knowingly. I had made an unauthorized export to Cuba without permission from the U.S. Department of Commerce and OFAC, a violation of the embargo.

Stateside, I decided the best course of action was to come clean right away. In official terms, I would make a "voluntary self-disclosure" to the Department of Commerce. Better to admit the infraction up front, I reasoned. A compliance officer paid us a visit. He handed me his business card, which was adorned with an oversized embossed gold badge, probably intended to intimidate. As we discussed my actions, I reminded him that this was a matter of trying to save lives. With no paucity of attitude, he proceeded to tell us that if it ever happened again, we would risk being able to continue our work in Cuba. He then went on, "Another part of our job is to protect national security. You guys may not think of scuba tanks to Cuba as a national security issue, but scuba tanks to an embargoed country makes it a national security issue." Thankfully we received only a slap on the wrist.

But our sigh of relief was short-lived. I was soon summoned to the State Department to explain my actions. Seated at one of the longest conference tables I had ever seen, and seated among people I did not know (and who had no intention of introducing themselves), I was interrogated and lambasted, one woman clearly resentful that I was working in Cuba at all and unreservedly telling me so. Thankfully, there would be no long-term consequences except a sour stomach that would persist for days. No good deed . . .

After writing about Alonso's death in our newsletter, it was touching to see donations for Alonso's widow coming to us from all over the country. On my next visit a month later, I gave her an envelope containing the donations. Tearfully, she shook her head in disbelief. "I can't believe so many people in the United States would care about me."

Dead Things in Jars

The unavoidable potholes on the road of Cuba–U.S. collaboration would fill volumes. And so, over the years, in the dozens of meetings at CIM that would follow that first meeting with Drs. Suarez and Ibarra, I met countless representatives of universities and NGOs visiting the center, eager to establish programs in Cuba. I would never see 99 percent of them again. I was certain that one taste of the challenges they would face was enough to change their minds about Cuba.

During that first trip in 2000, I learned about the work the Ocean Conservancy (where I was vice president) was doing in Cuba. Dr. Ibarra took me up to the second floor and showed me the scientific collections of fish and invertebrates, each specimen paled by years confined to a formaldehyde-filled jar. With Ocean Conservancy's help, new cabinets and preservation materials arrived to help rescue their collection, pains-takingly collected and cared for over decades. We later talked about future collaboration. Dead things in jars are important to science, but I explained that as vice president for conservation policy, my role was to advance the conservation part of our organization's mission statement. Science is certainly part of that, but science that advances our understanding of how to best conserve our marine environment. They smiled in understanding, and our discussion broadened to joint research, maybe expeditions, sci-entific papers, and information that could advance marine conservation. Blissfully unaware of the challenges ahead, we committed to a long-term relationship. A little more than a year later, a small group of Americans and Cubans would gather on a beach to create a plan for what would be an ambitious, decade-long project together.

PART FOUR

TIME TRAVEL

Regret and time travel are intrinsically linked.

—Colin Trevorrow

CHAPTER TWELVE
IN SEARCH OF CINDERELLA

What interests me about the life of an explorer is you are in the unknown; you are out of your habits.

—Bertrand Piccard, French explorer

Uncharted Waters

We gathered on a sandy beach on the small cay, Cayo Levisa, a tourist locale on Cuba's northern coast, west of Havana in 2003. Our team from Ocean Conservancy included, among others, scientists Dr. Cheri Recchia and Jack Sobel. We were also joined by board member Shari Sant Plummer, who would later create and lead the Code Blue Foundation, an important supporter of Ocean Doctor's work in Cuba. The Cuban team from CIM included Dr. Gaspar González and coral reef ecologist Dr. Elena de la Guardia. Joining the scientists, the lifeblood of our work—representatives from the foundations that made our work possible. Because of the economic embargo, funding from the U.S. government was impossible—and the Cubans would disqualify any project so funded.

It did not take Gaspar long to entertain us with one of his trademark stories. Learning that Cheri was Canadian, he recalled a cocktail party welcoming him to Halifax during a visit to Dalhousie University. "Suddenly all the men left the room—I think to smoke cigars—and left me alone with all the women." I imagined Gaspar, nervous and uncomfortable, sheepish grin on his face, a scientist accustomed to socializing with fellow scientists finding himself in the utterly alien world of cocktail party small talk, especially in a room full of women, especially in another country, in another culture, and in another language. He smiled and the women heard him say, "I feel myself the *bitch master!*" Gaspar recalls, "There was dead silence and no one was smiling." Unfortunately, Cubans have the unfortunate habit of pronouncing a long "e" in English as a short "i." So the "Beatles" become the "Bittles." And "beach" becomes . . . Gaspar was actually referring to himself as the "beach master," a moniker given to highly territorial male elephant seals who fight to control the beach and their harem of 20 to 100 females. In the highly unlikely chance that Gaspar's audience would have understood the intended meaning if correctly pronounced, it's almost a certainty that they would have been even more offended.

A common mistake among U.S. NGOs, academic institutions, and foundations is to come to a country like Cuba and purport to have all the answers, pushing their own agenda. It can be arrogant and condescending, dismissing the talent, hard work, and results achieved by the local scientists and conservationists. To fully appreciate the challenge and complexity of environmental issues and to really make a difference, there is no substitute for working locally, side by side with those who live there, as equals. Humility can work wonders to ensure trust, to build camaraderie and a cohesive team. And so, uncharacteristic for our chatty bunch, we shut up and listened to the Cubans. We simply asked, "How can we best support the work you're doing?" Gaspar pointed to the shoreline where small, gentle waves from the Gulf of Mexico quietly stroked the sand as the afternoon sun glittered upon them. "We know very little about these waters," referring to the 120-km stretch of Gulf from the tip of the wild Guanahacabibes Peninsula to the west, to the entrance to Havana Bay to the east. With the exception of a vague reference to a Soviet expedition that may or may not have occurred in the early sixties, and a fish-collecting expedition in 1914, these waters were unexplored, save for the areas CIM biologists and students could reach from shore by walking from the center. Offshore, only basic bathymetry data had been collected as the basis of nautical charts to aid navigation. Cuba's northwestern coast hugs the coast of three Cuban provinces: Havana, Artemisa, and the verdant Pinar del Río Province, home to Cuba's legendary cigars. Pinar del Río Province is the least-developed coastal region of Cuba. Unofficially referred to by some as the "Cinderella region" of Cuba, presumably because it had yet to find the riches part of its rags-to-riches story, I was initially enchanted by the epithet.

As an explorer, I found it tantalizing that ours would be among the first eyes to take in the mysteries of this region. But what made this multiyear journey more distinctive was the fact that we would be diving into waters unique from the rest of the Caribbean and the Gulf of Mexico, in the waters of an island larger than all of the other islands of the Greater Antilles combined. In addition, curiously, Cuba's coat of arms bears a prominent key in the center, symbolizing its self-proclaimed identity as *"llave del golfo,"* the key to the Gulf of Mexico—a subtropical nexus where the waters of the Caribbean Sea, the Gulf of Mexico, and the Atlantic Ocean intertwine in a sublime undersea cocktail of diversity, color, and mystery. To the east, Cuba's shelf is practically nonexistent. Motor a few miles offshore from Havana and you find yourself atop thousands of feet of water. To the west, however, the shelf is much broader around Banco San Antonio and the Archipiélago de los Colorados, a string of small cays running east to west along a barrier reef, with depths in much of the area between 20 to 30 feet, ideal for collecting data using scuba.

As a research institution, CIM's mission was research and not conservation per se. But the data the center collected would be invaluable to decision-makers at CITMA and Centro Nacional de Áreas Protegidas (CNAP), the National Center for Protected Areas, housed within CITMA. The data would serve as a foundation upon which to build conservation strategies. The goal was to create the first-ever maps of the region's ecosystems—ambitious to say the least.

The sun had fallen low in the sky, painting Cayo Levisa and its tourist bungalows in crimson and purple. Having completed our work, we silently gazed across the waters we would soon be exploring, now flat calm and rich with the color of the sun—our well-earned "psychic pay" for the day. The tranquil moment was interrupted by my noisy brain, processing the plan we had just constructed. Given resource, regulatory, and logistical constraints, it might take a decade to accomplish this. In fact, it would. But the results would be unprecedented.

Now, if only we had a boat.

You're Gonna Need a Bigger Boat

I traveled the width of the Cuban island south to see CIM's only vessel, the *Felipe Poey*, named for the revered Cuban zoologist of the 19th century, famous for his studies of fish, visiting Havana's fish market to obtain his samples. He founded Cuba's Museum of Natural History in 1839 and later became the University of Havana's first professor of zoology and participated in the creation of the Cuban Academy of Science. A painting of Poey at the Academy depicts him at the fish market, dressed impeccably in a black suit. I turned to Academy head Sergio Pastrana and remarked that this was no doubt an embellishment by the painter. "No, David. This is exactly how he dressed!"

Named for such an important figure, my high expectations for the vessel were dashed when I laid eyes upon a boat that seemed to be crumbling into the sea. There was something strange about the hull—painted black but pockmarked with small but deep white craters. "Our vessel is a Cuban innovation. It's made of stone," Gaspar told the journal *Nature*. The technical term is "ferro-cement," a fast and cheap way to create a boat hull, a common practice during World War II and proof that cement can, indeed, float. I boarded for the tour. A cockroach scurried by my feet. The boat was infested with them. The vessel had no generator, just a 12-volt car battery to power the running lights. There was no navigation system or GPS. There was no air-conditioning, and it was stifling belowdecks. The crew, scientists, and students would pitch tents and sleep on the upper deck. Despite its lack of creature comforts, the *Felipe Poey* served CIM well in the calm waters of the southern coast. But I had great trepidation about using such a vessel in the often rough waters of the Gulf of Mexico. It turns out that the Cubans agreed. At some point I had heard that CIM had a second vessel and wondered if that might be more seaworthy. I was told they did indeed have a second vessel, but for the past five years it had been residing at the bottom of a river.

Our first task, then, was to find a suitable, seaworthy vessel. I had heard about a converted Soviet fishing vessel called the *Ulises* and was referred to Paulina Zelitsky, a Ukranian-born engineer living in Canada and working in Cuba, where she believed she had discovered, incredibly, a sunken city 2,000 feet underwater off Cuba's northwestern coast. She was using the vessel for her exploration work and agreed to give me a tour.

She picked me up and drove me to the marina. The vessel was massive, 10 times too large for our needs. A trawler, it was probably used to fish just outside the 12-mile territorial limit of the United States before Ronald Reagan extended that limit to 200 miles. Exiting the car, she said with urgency, "Put your head down!" My first thought was that someone was going to shoot at us. I then realized, to my horror, that we were trying to evade the gaze of security. Paulina hadn't obtained official permission to take me aboard. We scurried, heads down, and managed to board the ship. Our tour—which included a 30-person sauna—was suddenly cut short on the bridge and Paulina guided me into a storage locker. "Stay here and be quiet," she instructed me. A security guard was making his regular rounds. Cubans aboard boats are always a great concern for the Cuban government, given the steady stream of Cubans trying to make the journey across the Florida Straits to the United States. It's hard to say what sort of reaction the presence of an American would have caused had I been discovered. Thankfully, my brief visit to the *Ulises* remained a secret (until now).

Paulina invited me to her Cuban residence to see the side-scan sonar images of the "underwater city." Highly skeptical at first, the images were jaw-dropping. There appeared to be a long promenade, adorned on each side by perfectly symmetrical columns. There appeared to be an amphitheater, again, perfectly symmetrical. A number of other structures looked man-made. It was like a bird's-eye view of an ancient Greek city. Or perhaps the lost city of Atlantis? As much as I wanted it to be true, it seemed rather unlikely that Atlantis would find itself in Cuban waters, and it was way too deep, even accounting for sea level changes. Still, it was a gripping mystery. Nearly 20 years later, they would conclude that they had discovered a remarkable series of rock formations, but alas, no ancient city.

The best vessels in Cuba are reserved for tourists. So María Elena and I decided to venture to the local offices of Ecotur, which runs all of Cuba's ecotourism enterprises. We were warmly welcomed and, in Cuban tradition, served espresso. We explained our needs for a vessel as they took notes. As we were close to wrapping up, a question came out of left field. "Open bar?" the director asked. It was one of those English terms that has made its way to Spanish unchanged. I had to explain the concept to María Elena. "No, no, you don't understand. This isn't for tourism; this is for research," I explained. Perhaps the fact that I was American threw them, and they didn't take warmly to the fact that Cubans would be aboard. They put together an invoice for a tourist excursion that was prohibitively expensive, even by American standards, and even without an open bar. We retreated to CIM empty-handed.

There were precious few suitable boats in Cuba. Back at CIM, María Elena, Gaspar, and I pondered our next move. Bringing a vessel in from the United States was possible but would require complicated licensing, months of waiting, and the probability that the license application would be rejected. These were the years of the George W. Bush administration and relations between the two countries were decidedly on the chilly side. María Elena and Gaspar were aware of a vessel called *Boca del Toro* (Mouth of the Bull), officially the vessel of Flora y Fauna, a Cuban agency not unlike the U.S. Fish and Wildlife Service, responsible for managing natural resources and enforcement of regulations. Would they be willing to lease the vessel to CIM? The only way to find out would be to hold a meeting with the head of Flora y Fauna, and that was an intimidating proposition. The agency was headed by Guillermo García Frías, at that time one of only six Cubans with the title Comandante de la Revolución, a title reserved for those who fought at Fidel Castro's side during the revolution. To say I was unnerved is an understatement. And that was before María Elena insisted that I lead the meeting.

With the meeting arranged, I decided it would be best to have an interpreter present. Though I would eventually become near-fluent in Spanish, at that time my Spanish was weak and still polluted by my Russian. I just wanted to be sure I didn't inadvertently create an international incident. My go-to interpreter was unavailable so I was referred to a young man who would serve in her place. Flora y Fauna is housed in a nondescript, unmarked building in the Vedado section of Havana. María Elena and I exited CIM's Lada, entered the building, and were led by two army guards—with sidearms—to the conference room where the comandante awaited. He was in full military garb, including his hat, and the "fruit salad" on his chest seemed to reach down to his belt, more ribbons than I have ever seen adorning any military officer. He was short in stature, stocky, and looked at me with dark, piercing eyes. He was intimidating as hell and I think he knew it. I was relieved when he offered a warm greeting, a welcoming handshake, and even a nascent smile.

Comandante Frías took his place at the head of the table. I sat closest to him, María Elena next to me. My interpreter sat against the wall behind me. Several other representatives of Flora y Fauna sat across from us. After sipping our espresso, we were officially welcomed by the comandante, speaking slowly and deliberately, after which I felt the burn of expectant eyes upon me. Speaking in English, I thanked the comandante and his team for their hospitality and willingness to meet with us. I explained who I was, the organization I represented, and why collaboration between Cuban and U.S. scientists was so important. As I was about to launch into the specifics of our request, I became aware that I could barely hear our interpreter. In *Seinfeld* parlance, he was a "low talker." If I couldn't hear him and he was directly behind me, I couldn't imagine how anyone else in the room could hear him. As I ventured further into my dialogue, I also realized—even with my limited Spanish—that his interpretation wasn't even accurate. At that moment, to my horror, I faced the inevitable: I was going to have to dig deep and switch to Spanish. And so, with a deep breath, I did. I spoke for about 90 seconds, and the comandante suddenly raised his hand for me to stop. I felt sweat gathering under my shirt. Had he not understood? Had I butchered the language so badly that I had insulted him?

He stared me in the eyes. I waited for what seemed an eternity. Curiosity painted his face as he asked, "*Como es posible que un gringo de Washington, DC, este en mi oficina hablando español perfectamente?* (How is it possible that a gringo from Washington, DC, is in my office speaking Spanish perfectly?)" I was in disbelief. And overjoyed. I continued in Spanish and stumbled through my reply to his question, saying something to the effect of, "Well, sir, I am here working in *your* country. Speaking Spanish shows respect to you, my colleagues, and the Cuban people, and it helps me better understand your culture. I think doing our work here is impossible without that understanding." Whatever I said caused him to beam with a warm smile. We were the unlikeliest of allies from that day forward. And the team from CIM and I would soon set sail on the *Boca del Toro*.

Where Are We Going?

We were euphoric. We had a boat. Now we needed to know where we were going. CIM had no GPS, which was illegal to bring to Cuba without an export license from the United States, and illegal by Cuba's own regulations. Strangely, I lost my personal GPS unit, only to find it a decade later in Gaspar's desk drawer in his office at CIM. Well, at least that's the official story. It's one thing to know your latitude and longitude, but you also need a map and CIM had no suitable nautical charts. Unsure of the accuracy of a U.S. nautical chart, the most reliable chart, I thought, would be one created in Cuba. Asking around, I was told that there was a nautical store in Old Havana. "Just look for the ship's wheel," I was told by a colleague. And so on a sweltering afternoon, I combed Old Havana until, sure enough, I spotted the ship's wheel beneath which was a nautical shop, a bizarre sight in this touristic part of town full of small state-run restaurants, gift stores, and galleries. I entered and immediately found the charts we were looking for. I purchased two sets, one for Gaspar and one for me. The proprietor rolled them up and stuffed them into two cardboard tubes. Gaspar was delighted. Now we could plan the sites we would visit for our research, knowing with sufficient precision their latitude/longitude coordinates and water depth. I would return home to Washington, DC, to study the maps and then confer with Gaspar.

Having cleared passport control, I placed the tube of charts on the X-ray belt. A security guard asked me, "What is in the tube?" Before thinking, I replied, "Those are nautical charts of your northern coast . . ." suddenly realizing that, from the security guard's perspective, my sentence might likely have continued, ". . . to plan the U.S. invasion of your country." I was quickly escorted to an interrogation room. Two young Canadians shared the room with me, busted for trying to return home with black market cigars—easily obtained on the street but easily detected by security. Two security guards sat me down, polite and even apologetic, but clearly serious and, at a time when very few Americans visited Cuba other than Cuban American families, intensely curious about this *Norteamericano*. I did my best to explain the situation, one that I'm sure sounded like fantasy—Americans and Cubans working together to sail off into the blue and scuba dive on unexplored coral reefs. Getting nowhere and now nervous about missing my flight, I pulled a wad of Cuban business cards out of my pocket. Like a card game, I held the stack with my right hand while with my left, laid each card on the table, scanning their faces for a look of recognition. Nothing. I flipped through half the deck until, at last, they looked at each other and nodded their heads. I looked down. It was María Elena's business card, my get-out-of-jail-free card. The guards politely apologized and I was soon on my way—with the charts under my arm and no plans for an invasion.

Beware Cinderella

As the project began to come together, it fell on the U.S. team to raise the funds to make it happen. A catchy title of a project is a business tradition in the Western world, whether nonprofit or for-profit. Still enchanted with the name "Cinderella" used to describe the Pinar del Río Province, we settled on "The Cinderella Project." We loved it and so did the donors. We used it in presentations, proposals, and it even appeared in a news-paper article. We successfully raised the needed funds that would buy the diesel, rice, beans, and other provisions, along with some equipment we could legally (and easily) export.

During my next visit to Cuba, preparing for a meeting, I was abruptly pulled aside by a clearly agitated Sergio Pastrana. "David, you *have* to change the project name!" I was bewildered, not only by his request but also by his near-panicked demeanor. "The people in Pinar del Río Prov-ince are very sensitive. There are a thousand jokes about them." I still didn't understand the connection. Seeing that I was clueless, Sergio cut to the chase. "Using 'Cinderella' is like saying 'nigger' in your country!" I stood dumbfounded—and ashamed. I hadn't had a clue, and I realized from that day forward that "catchy" names were not the way in Cuba. In fact, Cuba has the distinction of some of the most boring project names imaginable. But in a country where you don't want to draw attention or risk political incorrectness, it makes perfect sense. During our meeting with Gaspar and María Elena, the project was given a title makeover and anointed with the hopelessly unimaginative but politically correct title Proyecto Costa Noroccidental (Project of the Northwestern Coast), an international crisis avoided.

Counting Benjamins

Even with such a boring name, we were able to raise funds for the project—for a decade. But given the economic embargo and no formal banking relations between Cuba and the United States, how could I get the funds to Cuba to pay for the lease of the vessel, food, ground transportation, and provisions? We had obtained a license from OFAC allowing us to make the expenditures for these items in Cuba, but getting those funds to the island was quite another story. In the end, we concluded that the most reliable and easiest option available to us was to carry cash to Cuba. In fact, this was tradition on all of our trips. U.S. credit cards and ATM cards didn't work in Cuba, so we carried cash and hoped we wouldn't run out before the end of our trip. On occasion, I did run out of cash and had to borrow from friends and colleagues.

To fund the expedition, I prepared myself to carry $15,000 in cash—150 crisp Benjamins—in my pants. To keep the bills secure, I used a concealed money pouch that hung inside my pants by a loop around my belt. What I didn't anticipate is the request to remove my belt while entering Cuba and place it with my carry-on luggage on the conveyor to be X-rayed. With my belt removed, the pouch began to slide down my leg inside my pants. I placed my hand just above the knee to catch it from falling, putting my other hand on my hip, smiled, and tried to appear casual and inconspicuous as I awkwardly but somehow convincingly shuffled my way through the inspection area.

I would learn the next day that I had only cleared the first hurdle. At CIM I was informed that we would need to officially register the cash with the university's bank before it could be expended. Coqui fired up her Lada and the three of us—Coqui, my money belt, and I—were off to the bank. We were greeted by a flustered bank official and a stack of paperwork to complete. But first, the official needed to record, by hand, the serial numbers of all 150 bills I had brought with me. The slightest imperfection—a stray pen mark or torn corner—would result in the bill being declared worthless. Thankfully, the bills were new and flawless. However, the serial numbers were not consecutive, so recording them was painfully time consuming. It took forever, followed by inspection of my passport and the paperwork I had completed, which included questions about the origins of the funds and assurances that they did not come from illicit trade in drugs, or worse, the U.S. government. It took the entire afternoon before finally receiving the blessing of the bank.

Departure

Back at CIM, Gaspar and Coqui briefed me on the final plan. Our research team, including six graduate students who had never been farther from shore than they could swim, would concentrate on Los Colorados and cover as much ground—err, water—as possible. To achieve this, we would be using a "rapid assessment" approach. Sampling stations were identified in advance and marked on the chart. In teams of one or two, we would pull up at a station, leap into the water to conduct transect studies on the corals—to identify species, health, and cover; examine and count invertebrates; conduct fish counts; and collect algae. Topside we'd use water sampling devices to collect physical data, including depth, temperature, and conductivity (salinity). We would deploy a Secchi disk, a round white disk with black markings lowered into the water column until it can no longer be seen. The "Secchi depth" is an indicator of water turbidity, often used as an indirect way of measuring the nutrient load of the water. Fewer nutrients means less plankton, which, in turn, means clearer water. We would also do water sampling for ecotoxicological analysis, to assess land-based pollution, a study Coqui was leading. The plan was to spend 15 minutes at each station and quickly move on to the next. In Cuban time, that meant we would actually spend 30 minutes at each station.

The *Boca del Toro* awaited us in La Coloma, 120 miles from Havana on Cuba's southwestern coast in the Pinar del Río Province, the region's primary commercial fishing port. Coqui and the rest of the students and crew awaited us as Gaspar and I loaded our gear aboard the boat, dodging the plantains hanging from the ceiling over the deck, ripening in the afternoon sun. I was introduced to the students I had not met, including a young woman named Oyaima (Oh-yah-EE-mah). It would take me weeks to get that right. By then I was accustomed to the fact that it was almost impossible to remember Cubans' names. There were plenty of Miguels, Jorges, and Marías, of course, but there were equally dozens of made-up names, many beginning with the letter "Y," such as Yaniel, Yolexis, Yoanni, Yunior, Yusnel, Yordanis, Yurisbel, Yanisleidi, and Yureidy. I've met several people named "Usnavi (Oos-NAHV-ee)," which is said to originate from Cubans living near the U.S. Guantanamo Navy Base. Painted on each building: "U.S. Navy."

It was a tranquil afternoon in the quiet harbor. Most of the fishing vessels had returned, secure in their slips. The smell of diesel was thick in the air as our tanks were being topped off. Little by little, fishermen across the channel, enjoying an afternoon slash of rum, gathered to see this strange vessel, crew, and equipment in their harbor, not to mention a clearly out-of-place American. Fueling completed, the engines roared to life, and the captain signaled us to cast off the lines. Free and clear, we began to leave the harbor and the curious onlookers behind. Walking aft to enter through the stern hatch, I gave a polite nod to the fishermen on the dock. Too late did I realize I was stepping into a huge slick of diesel fuel carelessly left on the deck after our refueling. I saw my feet pointed skyward as I fell hard on my American ass. My unusual nautical maneuver appeared to be a hit with the audience.

As maddening as the preparations and logistics had been to prepare for this expedition, there was one last detail I needed to review with Gaspar. This was to be a two-week expedition, but Gaspar and I could only stay aboard for one week because of other commitments in Havana. Gaspar explained that he had arranged for a taxi to drive more than 100 miles from Havana to meet us at a tiny, obscure harbor on the northern coast. It seemed so absurd I couldn't protest. Or perhaps I was too weary from all the other hurdles we had cleared. "Sounds great, Gaspar!" He reminded me to have cash ready to pay the driver—and tip him well.

We headed west into the sun, which lingered above the horizon until we cleared the long, narrow, and gorgeous Guanahacabibes Peninsula, then turned north toward the Gulf of Mexico through the 150-mile-wide channel separating the peninsula from Cancún. We were now sailing through the famed "key to the Gulf," the passage through which powerful warm currents from the Caribbean to the south push northward, the headwaters of the great Gulf Stream, which would race along the East Coast of the United States, past Bermuda, and then cross the Atlantic, delivering its warmth to the United Kingdom. We gathered in the galley to enjoy the sunset, already painting the port side of the *Boca del Toro* with its gentle orange light. I looked around at the graduate students excitedly taking in the view on their first expedition. Until that point, I had been so focused on planning a scientific expedition, it hadn't occurred to me that this effort went well beyond the study itself. I was looking at Cuba's next generation of marine scientists. This expedition—and those that would follow—would provide them with the basis of their graduate studies. Indeed, dozens of scientific papers would be published in the years ahead. We sat down to enjoy our first dinner together aboard the comandante's boat. The sun disappeared and a brilliant tapestry of stars emerged from the dark sky. A warm breeze poured through the open hatch as we enjoyed snapper, rice, and beans, giddy with anticipation of our first dive the next day. After so much preparation, we were at last under way, content in the magic of the moment. I eyed a bottle of rum stashed among a box of provisions—I had no doubt it would be empty before the evening was over.

Siren's Call

At sunrise, I was finishing my second espresso as I heard the anchor chain rattle on its way to the bottom. The engine went silent for the first time since we left La Coloma. We had arrived at our first station. It was glassy calm, so preparations were easy as we assembled our scuba kits and prepared the scientific gear. I gazed northeast toward where the Florida Keys lie on the other side of the Florida Straits. In the late nineties, my daughter, Anna, went to Seacamp. She was the same age I had been when I first attended. I hadn't visited Looe Key for years and took the opportunity to visit. I realized that her young eyes were seeing a pale shadow of the miracle of nature I had beheld. Some of the corals were bleached bone white; others were shackled in diseased bands of black. Many more lie smothered in broad blankets of algal slime that had robbed the reef of its rainbow of colors, leaving a lifeless green-gray skeleton where countless diverse fish and invertebrates once peeked from every crack and crevice. I haven't returned since.

Ready to dive, I back-rolled into the water. The first mate handed me my video camera, and I descended into the embrace of 30 feet of clear, warm water. In a Caribbean coral ecosystem, the massive, stony reef-building corals capture the most attention, built over thousands of years from the calcium carbonate left behind by generations of corals. Such reefs can be mountainous and full of fish. That's where you'll invariably find the touristic dive boats. It's also where you'll often find fishing boats. Countless fish depend on the reef for food and protection, their movement about the reef a kaleidoscope of color and movement. But our first station was not a massive reef. Almost ignored by tourists are large expanses with little stony coral and few large reefs, but adorned with "soft corals," gorgonians, coral colonies that grow as tree-like structures from a holdfast on the bottom. Like their stony cousins, they host symbiotic algae that impart bright reds, oranges, and purples to their thick branches. Among them is the common purple sea fan, a striking soft coral resembling a fan with delicate latticework of calcite as if handwoven and dyed in bright purple. They are oriented so that the flat end of the fan faces the current, the tiny polyps of the colony eagerly plucking plankton from the water. It is a delicate beauty, the sea fans gently waving along with the other soft corals as if summoning us to visit, a siren's call. Sponges and seaweeds sprout from the bottom, and the occasional stony coral head, like beige spheres of brain coral, dot the seascape. Small fish dart about. Compared to the frenetic motion and color of a large reef, it is a more subtle beauty. That morning it was nothing less than enchanting. The corals were healthy. The waters were clear. The algae were in check. It was a scene I had not witnessed for years. But there was little time to take in the scenery as we divided into our teams and began the assessment.

Vertical in the water column, Gaspar rotated his body and recorded each fish he saw using the "Bohnsack method" (yes, named for Jim Bohnsack). He scrawled on his underwater slate and pencil. Two graduate students went after algae. Graduate students Abel Valdivia and Patricia "Patri" González focused on corals. Patri had a love affair with coral reefs that long predated her graduate studies. "I grew up very close to the sea—just two blocks away—in a very small town close to Havana. Almost daily my grandpa took me to the sea early in the morning and then again in the afternoon. I learned to swim when I was three years old, and I spent all of my vacation there," she told me. She visited the beach year-round and was intrigued by the changes she saw. Her young, curious mind had so many questions. "How did the sand disappear into the sea during winter season and come back during the summer season? What happens with the animals that I can't see sometimes like the crabs when they enter into the sand, or where do the fishes go when I can't see them? So I grew up with many of these kinds of questions." Her uncle was also one of her guides to the wonders of the sea. She recalled one day when he caught an octopus:

"I observed the behavior of how they change the color of their skin, how they enter in the small caves in the reefs. And I learned with my uncle to observe very carefully all the animals inhabiting the reef." In other words, Patri has mastered the art of "looking small." When she was 15 or 16, she already knew she wanted to be a marine biologist and focused on being admitted to the University of Havana and voraciously consumed book upon book relating to the subject. A few years after the expedition, I received a joyous email from Patri. She attached a photo of herself, now "*Doctora* González" proudly standing at CIM having just defended her thesis a few hours earlier. Just a few years later she would become the director of CIM.

Topside a few minutes later, it was all smiles. We had gotten our first glimpse of Los Colorados, and it was a beautiful one. And we had our first data in the can.

Radio Play

My eyes traced from Gaspar's fingertip out to the horizon where he was pointing as he shouted to the captain on the bridge above. Barely visible, it was the first boat we had seen on the expedition. We motored toward it and as it came into view, I could see its hull was a mosaic of rust, tiny blotches of blue, the only meager evidence that it had once been painted. It was a fishing vessel, and its crew didn't look much better than the boat. Shirtless, they were drawn, dried, and wrinkled by the sun. Rope hammocks seemed to be the only amenity on a vessel that appeared simply miserable. Fishing poles were the only gear aboard.

As we approached, Gaspar hailed the vessel by radio. After exchanging greetings, Gaspar began a dialogue that would be as much performance as information exchange. "*Buenos días*! We happen to have some extra diesel and rum. Could you use that?" The radio crackled with an immediate response, "Yes, we could, thank you very much!" After a short pause, the fishing vessel radioed, "We happen to have some extra fish—could you use them?" Gaspar gives me a knowing grin. "Yes, of course!" he shouts into the microphone. We pulled alongside and made the exchange, a maritime example of how the wheels of Cuba's economy are greased. An outright trade would be illegal, hence, the careful wording of this radio playlet lest anyone be in earshot of the transmission. A bin of large, fresh snapper was hauled aboard. But the fish were intercepted before making it to the galley. The culprit, of course, was a scientist. I found Gaspar, measuring tape in hand, in the grand tradition of Felípe Poey, methodically measuring each fish.

Taxi!

A few days later, I stuffed a Cuban flag and an American flag I had brought from the United States into my vest before entering the water. Underwater, Gaspar and I, floating side by side, held the flags as Abel snapped a photo, one that I still use today to symbolize our collaboration. The teams then descended on a stunning reef, healthier than any I had seen in the Caribbean in years. This was the hard coral, stony reef-building corals, and the reef, while not massive, was vibrant and healthy. There was practically no disease evident. I spent most of my time near the surface at the reef crest with Abel where beautiful, healthy young elkhorn coral was colonizing. It was here that he snapped some of the most beautiful images from the expedition.

Back aboard the *Boca del Toro*, Gaspar took me aside, playfully scolding me that I had almost caused an international incident. "The first mate saw you with the American flag. He thought you were going to put it up the mast! The comandante's vessel. Imagine!" We laughed.

The week went quickly and it was time for Gaspar and I to disembark. We turned to the south to locate the tiny harbor, our rendezvous point with the taxi from Havana. I could see that even Gaspar had his doubts. We shaded our eyes with cupped hands as we peered off the starboard side of the bow, but all we saw was water and mangroves. Others assembled on deck to join in the reconnaissance. Someone shouted, "I see it!" Someone with younger eyes—neither Gaspar nor I could see anything. Then, all of a sudden, a tiny yellow dot was barely visible in the distance. Sure enough, as we approached, a small yellow taxi was parked in the shade of a mural of Che Guevara that adorned the side of an old building. Its driver, appearing hopelessly out of place in his neatly pressed uniform, awaited us dockside. On our journey back to Havana, we talked about the beauty of Los Colorados, its patchwork of soft corals and mosaic of brain corals, brilliantly colorful in the warm, clear waters, showing little disease. It appeared healthy, but we agreed and the data confirmed: There was something missing. We saw almost no large fish.

CHAPTER THIRTEEN

DON'T SAY "HARVEST"

A fish is more valuable swimming in the sea maintaining the integrity of oceanic ecosystems than it is on anyone's plate.

—Paul Watson

Pooping Coral

You would expect a landmark report on the fate of coral reefs in the Caribbean to bear the image of beautiful Caribbean corals on the cover. Not if Dr. Jeremy Jackson, emeritus professor at Scripps Institution of Oceanography and senior scientist emeritus at Smithsonian Institution, has anything to do with it. He edited and led in the development of the report. Dominating the cover of "Status and Trends of Caribbean Coral Reefs: 1970–2012," is a beautiful illustration of a colorful stoplight parrotfish, but its beauty is overshadowed by the fact that it is helplessly caught in a gill net. In the background, a pale gray degraded reef is visible. A lonely, dead stand of brain coral sits alone among the rubble of its former coralline neighbors.

Jeremy and Nancy Knowlton are husband and wife, an underwater power couple. Nancy jokes that in terms of the messages they convey about coral reefs, Jeremy is "Dr. Doom and Gloom" while she offers a brighter perspective and reasons for hope. Beneath Jeremy's mop of long, tangled blond hair reaching his shoulders and the earrings dangling from his ears, comes a voice of clarity and unapologetic bluntness rare among

153

scientists. Jeremy was a pioneer in being outspoken in public fora about the problems facing the oceans. I recall at Ocean Conservancy in the early 2000s we tried desperately to get more scientists to speak out, using Jeremy as an example. Most scientists felt that this simply wasn't part of their job description. Publishing papers and making data available for loudmouthed advocates was the extent of their duties. During that time, I recall a particularly heated argument I had with a scientist at Conservation International when we were discussing our joint work on the well-known IUCN Red List as members of IUCN's Species Survival Commission. He argued that our job was to produce the report, period. I argued that our role isn't just to toss the report onto someone's doorstep and walk away. We should ring the bell, put our foot in the door, invite ourselves in for tea, sit down at the kitchen table, open the report, and *explain* the Red List to them. Such arguments continue to the present.

Speaking science to the public is difficult—requiring a completely different vocabulary, and the public is intolerant of equivocation so common in scientist parlance, frustrated by such phrases as, "the data *may suggest* that . . ." and the like. Muddied communications from authorities during the COVID-19 pandemic, especially in its early days, are a perfect example. Though improving in recent years, scientists "popularizing" science also run the risk of drawing disdain from fellow scientists and, at worst, being essentially ostracized from the scientific community. Astronomer Carl Sagan's groundbreaking role in the PBS series *Cosmos* came with tremendous acclaim but at the cost of undermining his relationship with a number of colleagues. In the end, he unapologetically enlightened millions of us.

I had known Jeremy casually for a number of years, but had never collaborated until we invited him to join us in Cuba. He sat down with Ximena Escovar Fadul, project manager at Ocean Doctor, and me as we described our work in Cuba. After taking it all in, he said, "You guys need a million dollars!" Who was I to disagree?

When it comes to pondering the health of corals, Jeremy's mind is on fish, explaining the cover of the report. Reading the first few sentences of the abstract of his 2001 paper in the journal *Science*, makes his focus on fish quite clear: "Ecological extinction caused by overfishing precedes *all other pervasive human disturbance* to coastal ecosystems, including pollution, degradation of water quality, and anthropogenic climate change."

He argues that we have forgotten what the oceans looked like before humanity began its massive extraction of fish and other species. The paper continues, "Historical abundances of large consumer species were fantastically large in comparison with recent observations." The paper, and a large body of his work and public speeches, emphasizes three things: Fish play a critical job in keeping coral reefs healthy; we've taken and continue to take too many fish from the sea; and we have lost perspective of what fish (and other species) populations were in the past. But Dr. Doom and Gloom actually offers hope. By understanding the underlying causes of ecological impacts—especially overfishing—restoration of marine ecosystems is still achievable by better management. In my speeches and writings, I try to expound on Jeremy's work by explaining that fish are not simply crops, like corn, that we "harvest" from the sea. Fish are part of the fabric of the ecosystem—they have jobs to do—and one of those jobs is to keep coral reefs clean of the algae that could otherwise smother them. Yet the basis of much of managing fisheries relies on the same equations we use for corn and other crops: maximum sustainable yield (MSY). Jim Bohnsack is lead author of a 2021 *Fishery Bulletin* article, at long last appropriately questioning use of the term "harvest." The article opens with the fitting quote, "The fish harvest is not a harvest produced by human labour. It is Nature's gift—some would say God's gift—to humanity," eloquently stated by J. K. Archer to the New Zealand Parliament in 1944.

So why a parrotfish on the cover? Ninety percent of a parrotfish's day is spent eating coral reefs—mostly the dead stuff. Parrotfish have hard parrot-like beaks along with bright parrot-like coloration and owe their name to the fact that they look exactly like what you'd imagine the underwater version of a parrot would look like, albeit decidedly slimier—and they don't talk. But they do make quite an underwater racket. Divers and snorkelers hear them incessantly and at great distance biting, scraping, and chewing coral. It is that hard beak that allows them to munch on the stony corals that make up the reef. It is common to see parrotfish swimming about, nonchalantly emitting a contrail of white "sand" from their anus, the end product of coral as it is pulverized passing through their digestive system. That sandy fish poop makes up a substantial part of the sandy halo surrounding reefs and coral heads. A single parrotfish can deposit an impressive 100 to 1,000 pounds of the stuff per year. So, fair warning if you find yourself scuba diving and kneeling in the sand near a coral reef. It may well have traveled through the gut—and anus—of a parrotfish to get there. Feel honored.

On its face, a reef-eating fish wouldn't seem to be a good thing. But parrotfish aren't after coral—they're after the algae that grows on the coral, so they're quite selective as to where they point their beaks. In the end, the coral colonies find far more benefit than detriment from a healthy parrotfish population. In fact, with the loss of the *Diadema* urchin in the 1980s, a creature that also delighted in grazing algae from coral reefs, the parrotfish is one of the last defenses against algae that would otherwise smother the reef, now a widespread phenomenon throughout the Caribbean. The fact that parrotfish are considered by scientists as one of the few groups of fish that can help build resilience in coral reefs is borne out in Jeremy's work and that of his colleagues, demonstrating that the healthiest reefs in the Caribbean are those where large, healthy parrotfish populations thrive.

The challenge, predictably, is that many of us humans find parrotfish delicious. In Jamaica, among other islands, parrotfish is a treat, typically steamed or fried and served whole, or incorporated into fish stew. In the U.S. Virgin Islands, an enlightened fisherman and colleague, Nicky Martinez, regularly has small parrotfish in his boat box after a fishing trip. They're popular eating there. From Nicky's perspective, though, he didn't target them specifically. He felt that he had no choice but to fill his box with whatever he could find, as he's watched fish on the reef diminishing and large fish are all but absent. With the loss of fish on the shallow reefs, Nicky has had to dive deeper and deeper. When we met more than 15 years ago, his doctor was already begging him to give up scuba diving. By the age of 28, Nicky had contracted decompression sickness (the "bends") three times already. The resulting nitrogen bubbles in the bloodstream can affect the nervous system, cause chronic neurological problems, and result in death. For him, and other fisherman like him, parrotfish and other algae grazers need to help fill the box if he's to eke out a living as a fisherman and feed his children.

The Global Coral Reef Monitoring Network 2014 status report concluded that protecting grazers, like parrotfish, can help to restore coral reefs, recommending that Caribbean countries institute laws and regulations, including complete bans on fishing them. The International Coral Reef Initiative has made similar recommendations. Some Caribbean countries established regulations to this end, including Belize and Bonaire, Puerto Rico, the U.S. Virgin Islands, Barbuda, Guatemala, Honduras, Mexico, and Turks and Caicos. The Dominican Republic established a two-year ban beginning in 2017. But while some countries are trying to limit fishing on parrotfish, others are just now discovering the delicacy, like the Bahamas where the fishery has recently emerged and is growing.

Fortunately, to the benefit of Cuba's coral reefs, parrotfish are not a targeted species or popular in the country—it doesn't sit well on the Cuban palate. My friend, Vilma Albelay, was repulsed when I asked the question. "It tastes terrible! Oh my God, it's disgusting!" Her father, Ramón, a fisherman when he was in Cuba, recalls, "Some people would eat that fish, but they taste like urine." It seems that the Cuban and Jamaican palates don't agree. The Cuban government fed it to Vilma and her classmates when she was in lifeguard school in 1999 at the tail end of Cuba's Special Period. It was not a common part of their diet, but Cuban institutions were desperate for any protein that could pass as edible during those desperate times.

Beyond their critical role in maintaining the health of coral reefs, parrotfish are just plain cool and are a colorful delight to scuba divers. They regularly swap sex from female to male in a process called protogynous hermaphroditism, changing colors in the process. And if you go on a dive at night, you might see a sleeping parrotfish, tucked under a small ledge of coral, surrounded by a mucus cocoon it has secreted. It's an ingenious way of keeping the bedbugs from biting. The cocoon protects them from parasites by masking their scent and creating a physical barrier. If you're going to spend 90 percent of your waking hours eating algae and coral, you most definitely need your beauty sleep. So aside from their critical role protecting coral reefs, we mustn't forget that our undersea world is much richer and infinitely more interesting when teeming with its noise-making, color-displaying, coral-crunching, mucous-sleeping, sex-changing, sand-pooping citizens.

We Got It Wrong

Jeremy shared his personal history as a marine ecologist and perspectives on coral reefs with my graduate students at Johns Hopkins, who were fascinated and riveted throughout. What he would describe was new to them, and they're not alone. What Jeremy would observe about what keeps coral reefs healthy, while now far better understood by scientists, remains unknown or poorly understood by the public and policymakers. "I spent 13 years of my life trying to understand the way organisms compete for space on coral reefs. And that was fun and really interesting. But as I was doing that, I became more and more aware of the precariousness of the situation." Jeremy was around the oceans from his childhood while growing up in South Florida and began his graduate research in 1968. "During the time that I did my research for my PhD, nobody was talking about ocean conservation because nobody was worried about the oceans."

He recalls the growing awareness of the tropical rainforests in the late seventies. But nothing about the oceans. "We were essentially living in a fool's paradise, taking it for granted that the ecosystems we were studying were pretty natural." He recalls that they would notice that fish were more abundant in some places than others. "You didn't have to be a rocket scientist to know that that was almost certainly due to fishing. But nobody ever seriously thought that how abundant, say, the fish were on a reef, might be a determinant of the health of the reef."

At the time, Jeremy was studying the reefs at the Discovery Bay Marine Lab on the northern coast of Jamaica. The reefs of Discovery Bay were well known around the world. "They were the best-studied coral reefs in the world. People came from Australia to see how we were doing our research. Back then we were pioneers in a kind of new ecological approach to understanding coral reefs," Jeremy says. In 1980, Jeremy's world—and his perspectives on coral reefs—would change forever with an event that had its origins off the west coast of Africa on July 30, 1980. Within four days, a tropical wave born there was a major hurricane. The term "major" is an understatement. Hurricane Allen was a massive, powerful category 5 storm, one of the most powerful ever recorded. To this day, Allen still holds the record for the highest sustained winds ever recorded for a tropical cyclone in the Atlantic Basin, a devastating 190 mph. To Jeremy's horror, its eye passed nearly directly over Discovery Bay.

"When we woke up in the morning, we looked out and we saw islands where there had been coral . . . there were no coral left and we saw waves breaking out to sea that were taller than the 40- and 50-foot trees along the shore." Jeremy and the other scientists there spent the day after the storm in the water. "It was really weird. It smelled of death, it smelled of this slimy death of the coral, mucus and all that kind of stuff. I picked up the phone, which was miraculously working, and called NSF [National Science Foundation]." Having spent 13 months doing research on his thesis, Jeremy explained he had baseline data and convinced NSF to fund a team to, for the first time, document the extent of damage to a known coral reef. NSF agreed. It was the first time anyone had done it. "We documented what had happened and we published that in *Science*." It became an influential paper, the first comprehensive study examining what hurricanes do to corals.

But Jeremy's satisfaction with that research was short-lived. "Because we thought we were so smart, we predicted how the coral reefs were going to recover. We got it totally, completely wrong. And we were the best and the brightest in coral reef ecology," he said with a modicum of self-disparagement. Confident that the reef would recover, the question was how and over what time period? Soon, however, they would wonder not when the reef would recover but if the reef would recover.

"The reason we got it wrong was because we'd never thought about the fact that there weren't any fish left. They started to sort of come back. But algae were becoming disturbingly abundant." Over the previous years they had observed reduced abundance and size of parrotfish and other herbivorous fish due to fishing. They weren't there in sufficient numbers to eat the algae and keep it from smothering the reefs. Jeremy's "aha" moment was that, without those fish, any disturbance could push the reefs over the edge. Without a major disturbance, the reefs had been able to survive on a knife edge. Without the protection of the fish, they had become far more vulnerable than anyone would realize.

And then it got worse. On the heels of Hurricane Allen was the Great *Diadema* Die-Off. Reports were spreading by telephone throughout the Caribbean in the pre-email world. "I predicted that this thing, whatever it was, was going to hit Jamaica. And all the sea urchins are going to die in three or four days. I flew up to New York to get married to my wife, Nancy Knowlton, and when I came back all the *Diadema* are gone." He immediately returned to the water. "The reefs in Discovery Bay looked like they needed to shave. They had this fuzzy green stuff all over." Jeremy and his colleagues soon observed the reef covered with seaweed 12 inches high, and the small recruits of coral trying desperately to recover were being smothered. "There was what we call a phase shift, an utter transformation of what had been a coral reef into an algae-covered rock pile. And to this day, those reefs have never recovered." The combination of Hurricane Allen, the fact that fish populations had been dwindling over the years, and now the complete loss of another key algae grazer, the *Diadema* black sea urchin, amounted to, in the words of Don Levitan, a "death blow" to Discovery Bay. It also set into motion Jeremy's focus on the relationship of fish and corals throughout his career, hence the par-rotfish on the cover of the 2012 report.

"We got it wrong because we forgot that people had changed the rules of ecology under which coral reefs work. I mean, basically, the way you can think of it is like the three little pigs. Corals build their houses out of bricks, and seaweeds build their houses out of straw. Now bricks are really good when you have hurricanes. The calcified skeleton makes them stronger, protects them against storm damage." He points out that the algae growing on the corals attracts grazers, like large parrotfish. But corals pay a big price when you remove those grazers. Corals grow at 1/10th or 1/20th or 1/50th the rate of rapidly growing seaweeds. "So if you remove the Big Bad Wolf, the grazers that keep the seaweed in check, the seaweed grows dramatically faster than the corals. It overgrows them and smothers them and that's the end of the story."

Jeremy sighs as he recounts falling into a common trap among scien-tists, who, myself included, sometimes trot out the word "pristine" when describing healthy coral reefs. "We thought we were studying a natural system. But we weren't." He recalls that the best-studied coral reef in the world—and the interactions among it between fish, corals, and algae—was based on the assumption that they were studying a healthy reef. In fact, it wasn't a pristine reef—it had been overfished for years. It was a beautiful ecosystem of seemingly healthy corals, but was, in fact, teeter-ing at the edge of free fall. "There are still actually very few places in the Caribbean still not overgrown by seaweeds. I have dived on some of those because the fish were protected, either because the people were smart, or in many cases, because the people don't have the boats to get out there. And they're too poor like they are in Cuba. So the reason Cuba has much healthier reefs than most other places is poverty and the price of gasoline. Very simple."

Jeremy has spent much of his career since Discovery Bay trying des-perately to change the perspectives of other scientists to recognize that the "healthy" reefs we are studying are not pristine at all; instead, they have long been disturbed by humans.

"I got really tired of reading papers about pristine coral reefs. I actually was in a big confrontation in a meeting where people said, 'Yeah, yeah. Jeremy studied the Caribbean but we in Australia, our reefs are pristine.'" That's when Jeremy got pissed off. "I got so mad that I decided to give a plenary talk at the International Coral Reef Symposium in 1996 in Panama titled, 'Reefs Since Columbus.' " (It was required reading for my class.) "On Columbus's second voyage, the green turtles were so abundant, the ships could not sail through them. I found old hunting data and that there had been at least 35 million of them." Today it's estimated that there are, at most, 90,000 nesting females remaining. "That was just one species of turtles in the Caribbean. They tasted really good, which was their problem."

"I ended up by saying, 'Please, please do me a favor and never use the word 'pristine' again. Because it's bullshit." The room was silent. "I thought, 'Oh my God, I'm in trouble.'" And then I got this great applause, and it's like, 1,100 people or something. In the old days we used Kodachrome slides, and I had the pleasure that people would take little ink pens and cross out the word 'pristine' in the slides for their talks." He paused and appeared to scan the faces of each of the students in the class on the Zoom checkerboard. "I'm telling you those stories because that's scientists who screwed up, not the policymakers. I mean, you people are interested in policy. But these are the best and the brightest coral reef scientists who didn't know."

Scraping the Bottom of the Barrel

So where were the big fish in Los Colorados? Like the ones I saw in Veracrúz, they were largely absent. But the striking difference between Los Colorados and Veracrúz was the health of the corals. But were these corals the walking dead, without the protection of algae grazers, vulnerable to an invasion of algae against which they were nearly defenseless? Unfortunately, a number of Cuba's fish stocks are overfished. On the bright side, fishermen, like the ones we encountered, principally use hook and line so fishing in Cuba is highly selective and low impact to the bottom. Increasingly, however, we've seen Cuban fishing vessels spreading gill nets across channels, indiscriminately catching anything that comes along and becomes entangled, the nontarget species collectively known as "bycatch." It's not just fishing itself, but the way we fish that can also impact marine ecosystems. Bottom trawling, essentially dragging weighted nets across the bottom, is especially harmful. It is highly destructive to bottom dwellers, including corals. The trawl scars we had observed in the Bering Sea were just the tip of the iceberg. Globally each year, an area equivalent to more than half the size of the continental United States is trawled, accounting for a quarter of all marine landings. This destructive practice has attracted attention and concern for decades. In recent years, a new consequence of bottom trawling has been documented. Bottom trawling is responsible for the release of enormous quantities of greenhouse gases from the bottom sediments it disturbs.

Recognizing its devastating impacts on marine ecosystems, Cuba banned bottom trawling in 2003, save one area in the south where trawling remains permitted for shrimp. In contrast, Cuba's neighbors, Mexico and the United States, still trawl their waters. A striking photograph by Brian Skerry at National Geographic shows a bottom trawl scraping bottom in Mexican waters. Clearly, any soft corals, sponges, or pretty much anything else would be lucky to survive the trawl's passing. No doubt, coral reef communities have benefited from Cuba's decision to ban this practice.

Protecting Fish—and Everything Else

Cuba's fishing industry has had mixed results. Regulations exist, but an ongoing challenge is enforcing those regulations. And although we have learned about the crucial role of fish in keeping coral reef ecosystems happy and healthy, protecting fish alone isn't enough. The challenge becomes how to protect all of the fantastically complex relationships of a coral reef ecosystem. One approach is to roll out dozens of regulations regarding the capture of fish, lobster, and other marine organisms; establish quotas and seasons; require permits; etc. Managing and enforcing such a tangle of regulations has traditionally been fraught with ongoing regulatory changes, ambiguity, and the need for intense monitoring.

What if, like our national parks, we could establish marine parks, reserves, sanctuaries, etc., collectively known as marine protected areas

(MPAs) that restrict fishing and other activities? Such an approach simplifies management and enforcement and can help protect the integrity of that ecosystem by protecting all of its inhabitants simultaneously. MPAs have been slow to take off. In the United States, they have been, and still often are, vehemently opposed by fishermen, fearful of being locked out of their "right" to fish wherever they please. (We conservationists hold that it is a privilege, not a right, to fish in territorial waters. The oceans are a public trust that belong to all of us—the largest public trust of the United States.)

MPAs are now considered one of the most important tools in our conservation tool bag. But do MPAs actually work? We would soon see extraordinary living proof.

CHAPTER FOURTEEN

COLUMBUS AND THE QUEEN

Paradise is exactly like where you are right now . . . only much, much better.

—Laurie Anderson

Race to Jardines

The Cuban Coast Guard officials at our destination, Jucaro, a small fishing village, would soon pack up and return home to their families, decreeing it was too late for the *Reina* to leave port for the 50-mile journey offshore. We were hours late, stymied by typical Cuban obstacles, including trying to rent a pair of reliable cars for the journey from Havana, an ordeal that took fully half a day, punctuated by bursts of exhaustive negotiations, fists full of cash, and crosstown dashes to promised vehicles that didn't exist. We were also slowed by the inevitable breakdown of one of our "reliable" vehicles along the highway. Thanks to a pair of tie wraps and Cuban-inspired ingenuity—ironically performed by American NPR reporter and *Washington Post* writer Nick Miroff—we were able to jury-rig a repair.

The six-hour drive east along Cuba's major axis became increasingly harrowing with every minute as a determined sun slid steadily into the mountains behind us. My grip on the wheel tightened until my hands throbbed. In Cuba, darkness transforms a drive in the countryside from a mildly challenging journey into a nail-biting adventure, especially on this final stretch of highway leading into Ciego de Ávila Province.

As the pale gloaming gave way to darkness, oncoming headlights scattered across our filthy windshield, causing momentary night blindness, frustrating whatever hope I had of being sure I was in my lane. Such certainty would have provided a morsel of desperately needed comfort, though far from a guarantee of safety.

This stretch of the *autopista* had three narrow lanes, impossible to discern in the darkness. "Which direction is the middle lane?" I asked my Cuban friend and colleague, Dr. Fabián Pina, who lived somewhere beyond the far end of this highway. His matter-of-fact response confused and terrified me.

"It's both directions. It's a passing lane," he replied.

I protested, "You've got to be kidding!"

Eighteen-wheelers with blinding headlights barreled toward us in the center lane, seemingly inches away, their explosive wake—together with my own instinctive reaction to steer away—lurching our small station wagon toward the dirt shoulder, where invariably our headlights would reveal the rapidly approaching rear end of a horse and the unlighted cart it was pulling. So I'd veer to the left and hold my breath, hoping we weren't coming up on a slow-moving vintage 1950s Chevy without working taillights.

This maddening dance down the highway continued into the early evening as the *autopista* slowly gave way to the narrow two-lane secondary roads that wound through acres upon acres of sugarcane. It was dark when we arrived. A regular presence with us in Cuba since that memorable meeting with CIM on the beach, Shari Sant Plummer was part of our away team. At a conference back in Havana, after years of enticing us, Fabián finally won us over. He invited us to visit a place we had heard about for years—an enormous, spectacular reef system along Cuba's southern shore known as Jardines de la Reina, Gardens of the Queen. As a principal researcher at CIEC, Fabián was enchanted with the place, did his research there, and even named his daughter Regina (Queen). Reserved and understated, it's nearly impossible to mention Gardens of the Queen without mentioning Fabián. But now we were terribly late and nervous that the Coast Guard official had packed up and gone home. Past the sugarcane, the road turned to dirt and mud and we slowly passed through the center of the tiny village to a large gated harbor. We held our breath. The Coast Guard officer had waited. We had made it in time. The *Reina*, an

80-foot-long converted fishing vessel awaited us, engines already running. We quickly boarded and were unexpectedly handed mojitos by a bear of a mustachioed man named Noel López. I would eventually get to know him and his family. He's the human closest related to a fish I've ever met, at home in the water, a seasoned divemaster and award-winning underwater photographer. He knows the world underwater better than the one topside. Becoming a scuba diver in Cuba was difficult for Noel. He is a self-taught diver, having assembled his first scuba rig from a mix of old Russian parts and whatever else he could scrounge together. He had more than one scrape with the law. Though not against the law, officials were uncomfortable with the sight of Noel and his gear. On one occasion, an officer expressed alarm that Noel might use the equipment to scuba dive to the United States. The senior officer looked at the junior officer and shook his head, simply saying, "No." Noel was free to go, but on other occasions he would find himself detained.

We gladly accepted the mojitos—medicine to soothe our frayed wits after the harrowing drive, but as scientists, it seemed odd. I had a brief flashback to the offer of "open bar" during the visit María Elena and I paid to Ecotur. But the mojito made sense—this was a tourist boat, not a research vessel. Dinner awaited, adorned by freshly baked breads in the shape of crabs and fish. Before we knew it, we were under way. After our six-hour drive from Havana, we would now have a five-hour journey across the Gulf of Ana María to the chain of islets and the barrier reef, 50 miles offshore. Exhausted, we soon retired to our cabins.

The Shark Whisperer

We awoke anchored in calm, clear waters, just offshore a curved islet with a small sandy beach and fringed with thick mangroves. The islands were uninhabited, only the remnants of a very old fishing camp remained on one several miles away. After a quick breakfast and espresso, the *Reina* motored roughly 20 minutes for our first dive. The water was blue and inviting. It appeared deep—the bottom wasn't visible from the surface. We assembled our equipment, untangled our straps, and at last were ready. It was 2009, and Nick had tape rolling for an NPR broadcast he was planning. I was ready except for my fins, and as I sat on the dive platform and began to don my right fin, I saw them for the first time. Three . . . no four

. . . no eight silky sharks, five or six feet in length, bearing their beautiful deep brown color and long oversized caudal fin. Soon there were 15
sharks coming right up to the boat. I lost count. Maybe there were 30 by
the time Noel jumped in and beckoned us to follow. Fabián watched us
with some amusement. Shari and I were no strangers to sharks, but I had
previously been in the water with only two or three at a time. Leaping into
the middle of a swirling mass of 30 of them was another story. Taking
a deep breath, with my video camera in one hand and the other pressed
against my mask, I jumped in . . . and lived. It was magic. The sharks were
curious and made close passes, but there was not an inkling of aggression.
I was sure they were being fed—why else would they aggregate so quickly
by the boat? Shari and I found ourselves in the middle of this midwater
production, sharks above, below, and on all sides. They were magnificent
and I immediately began to film. We followed the anchor line 60 feet
down to a flat bottom, but the silkys didn't follow, preferring the sunny
waters near the surface. On the bottom, we saw a wall on one side that
plunged hundreds of feet down. On the other, a gentle slope toward the
surface. The bottom was adorned with bright pink vase sponges, colorful
soft corals, and outcrops of brain corals, all healthy and happy. And then
there were the fish. Gray angelfish, queen angelfish with their dark blue
crown, French angelfish, and sargeant majors, with their characteristic
black "miltary" stripes, hovered above the reef. Butterfly fish and damselfish cautiously peeked out from the corals. Colorful parrotfish of all sizes
surveyed the corals, stopping for an occasional beakful of algae. Jawfish
peered out of sandy holes on the bottom, disappearing as we approached.
Huge schools of grunt and porkfish with their bright yellow tails carpeted
the upward slope, motionless but for the gentle movement of the surge.

But this was just the opening act. While the silky sharks remained
above, an entirely new school of sharks—Caribbean reef sharks—
appeared around us. Larger and more stout, they were gray with white
bellies, the tips of their pectoral fins dipped in black ink. Like their cousins above, they, too, were curious, and it was sometimes startling when
one would suddenly appear in front of our faces. I continued to film and
found that I could get quite close. They were likely also fed to bring them
close for the tourists to enjoy, and had been somewhat habituated to humans. Then there were the grouper. Huge black grouper appeared, beautiful with their deep brown skin, white mottling, and enormous mouths.

Like Labradors, they curiously followed us about. They were joined by the smaller but still sizable Nassau grouper, lighter brown with beautiful thick white stripes. They are considered a threatened species throughout their range in the Caribbean. I was thrilled to see one on the island of Saba months earlier. Here there were dozens. I had never seen grouper this plentiful or this large. I then looked up and beheld the mother of all groupers, the Goliath grouper. This was a small one, perhaps 200 or 300 pounds. They can grow to nearly 700 pounds. It hovered, its brown and mustard mottling distinguishing it clearly from the other species. They are imposing and highly territorial. They're also considered a critically endangered species. So, unlike the decimated reefs throughout the Caribbean, this was like a Broadway show: "Coral Reefs: The Original Cast." As beautiful and healthy as the reefs were in Los Colorados along the northwestern coast, this was something at an entirely new level. Not only were the corals healthy, but the big fish were here—in numbers.

Shari and I reluctantly began our ascent and noticed something odd near the surface. Noel, sans his dive equipment save his Speedos, was bobbing at the surface in a sitting position. Across his legs lay a six-foot silky shark. It lay motionless as Noel casually petted the brown-skinned back of what I suppose you might call a lap shark? We watched as Noel continued to pet the animal affectionately. When he stopped, the shark instantly swam away and disappeared into the blue. Shari and I looked at each other and telepathically communicated, "We're outta here." We made for the boat. Shari boarded as I bobbed on the surface to find myself surrounded by silky sharks, realizing that they were attracted by the irresistible shininess of my unpainted aluminum video housing—a giant fishing lure. Too close for comfort, I had to use the camera to push one of the curious sharks back, but it was in vain. The shark was determined to have a little taste of my camera, and so it did. He took the front of the camera into his mouth as I struggled to pull it out. I was still rolling—the video would be fantastic, I thought. Typical photographer. The sharks were still calm and curious, but soon I wasn't, as the circle of sharks around me tightened. I pushed several more away with my camera. As if on cue, the Shark Whisperer himself, Noel, came to my rescue. With Jedi-like agility, he parted the sharks and cleared a path for me back to the *Reina*. Later I was disappointed to see that the video from inside the shark's mouth was completely black, but the audio of its crunching teeth was spectacular.

Already aboard, Fabián had a huge grin. He could see in our faces the joy of what we had just seen and he seemed pleased and proud. Of course, he also saw our perplexed look about Noel's strange relationship with the sharks. What was that all about? It's a phenomenon known as tonic immobility. For silky sharks, bending the tail or tickling the snout puts them into what is essentially a trance, a temporary paralysis. Shari and I hadn't seen Noel's hand bending the shark's tail. We would later learn that some divemasters would first grab the tail to immobilize the shark, then carefully place the snout into the upward-facing palm of their opposite hand and with their fingers, tickle the snout. They would then release the tail and balance a six-foot shark vertically, tail straight up, on their palm. It was a favorite of tourists and photographers. For our future trips, I asked them not to do this with our visitors, explaining that I thought it sent the wrong message about how humans and wildlife should interact, and even more important, I knew everyone would want to try it. Unfortunately, not all of the divemasters got the message, and I was right—everyone wanted to try it. Stupid shark tricks . . . sigh.

We were effusive with excitement talking with Fabián and Noel about what we had seen. But it wasn't until the next day that I would truly enter the living time machine. We anchored as shallow as we could and would have to snorkel the rest of the way into less than 10 feet of water. I was simply astonished at what I saw. An enormous stand of mature elkhorn coral spread as far as I could see to the west and east. It was healthy, near-flawless, with every ledge and hole packed to the gills (pun intended) with grunt and snapper, the schools moving back and forth as one with the gentle surge. As I would be on every visit to this area, I was mesmerized. These stands looked even better than the ones I recalled from my teen years in the Florida Keys. And the stands continued for 30 miles to the west—a fantastic barrier reef. On my frequent visits to this area, I would regularly encounter a trio of large spotted eagle rays (listed as "near threatened" on the red list), flying in formation along the reef, gently fading from sight. For years, I'd get a bit of stink eye from the visitors I was leading back at the boat for making them wait as I explored like a boy ignoring his parents' calls. I would encounter many other corals throughout the area, all healthy and vibrant, one of the most dramatic being pillar coral, which looks like it sounds. The coral extends toward the surface in parallel pillars, each draped with unusually long tentacles that sway with the current, plucking plankton from the water. Pillar coral is considered a threatened species, and it's increasingly rare. Noel led us to the largest stand I've ever seen—perhaps 20 feet tall. And it was extraordinarily healthy.

From my days in the Keys, I also remembered tarpon, an immensely popular fish for catch-and-release sportfishing. These primitive-looking fish are large. They can reach eight feet in length and 300 pounds, though most commonly they're less than half that size. They bear huge scales of polished silver and sport a prominent underbite. As a teen, I thrilled at seeing a school of six speed by as I was diving. In the deeper areas, under the rocky ledges, there were hundreds. They floated motionlessly, and allowed me to enter the shallow caves, and to approach and film them. Together we floated silently. I held my breath to keep my bubbles from spoiling the moment. It was pure blue tranquility.

This magical area was named for Queen Isabel by Christopher Columbus during his second voyage. As recorded by his son, "the nearer they came to Cuba the higher and more beautiful these islets were. Since it would have been useless and difficult to give a name to each one, the Admiral called them collectively El Jardin de la Reina." He went on with vivid descriptions of the area: "They also saw great numbers of small birds, which sang most sweetly, and the air that below from the land was so soft that they seemed to be in a rose garden full of the most delightful scents in the world." Ironically, the enchantment of Columbus and his crew came only from what they experienced above the water. He never experienced the gardens below.

Has Anyone Checked Anderson's Air?

When Anya Bourg, associate producer at 60 Minutes, called and told me that they wanted to do a piece about coral reefs, I was ecstatic. It had been difficult to interest the media in the plight of coral reefs, let alone any ocean issue. Anya felt that Cuba would be an ideal location and Anderson Cooper was the logical choice as correspondent. He is a diver, fit, and had already done a piece in South Africa diving with great white sharks. I met Anya and producer Andy Court in New York and we discussed the story and logistics. Given the number of sharks, Anya was concerned about Anderson's safety. She asked if we could keep Anderson at a depth of 60 feet or less and without sharks. So, of course, on Anderson's first dive, he was at close to 100 feet and surrounded by sharks.

With full face masks and underwater radios, we were able to communicate. I listened as Anderson did a stand-up, talking about the area as huge grouper and sharks swam by him. Ever the professional, he did take after take after take. I lost count of how many. Over and over, I heard him repeat variations of what would finally be aired: "I've been diving in many places all over the world, and I've never seen so many large fish like this grouper here. There are about six or seven Caribbean reef sharks like this one circling around. Scientists will tell you the presence of so many sharks and so many species of sharks is a sign of a very healthy reef." Some of the takes failed because a huge black grouper would swim in front of Anderson's face and just hang there. On other takes, the sharks didn't time their entrances well. At some point, I looked at my watch—he had been down a while and talking the whole time. At that moment, the urgent voice of underwater cinematographer Bill Mills interrupted, "Has anyone checked Anderson's air?" With that, Anderson was, at last, on his way to the surface and took the last breath of air the tank had to give just as he was about to hop back in the boat. Sorry, Anya.

Both Anderson and I were prematurely silver-haired, so the film crew gave us different colored masks so they could tell us apart underwater. The piece was well researched and carefully scripted (except for the interviewees' answers, of course). Andy would intensely follow along his notes as Anderson interviewed Fabián, divemaster Andrés Jimenez, and me. The result was a 12-minute piece that earned the Edward R. Murrow award. It was one of the most-watched *60 Minutes* pieces at the time with more than 19 million viewers tuning in. It helped open eyes about coral reefs while also telling a positive story of collaboration between the United States and Cuba, something all too rare in media stories about the two countries.

For years, we had feared becoming too visible in our work lest we draw fire from anti-Castro Cuban Americans. We were shy when it came to media. And I worried about the reaction of 19 million people tuning in that Sunday night. I awoke the next day to hundreds of emails, including many from the Miami Cuban American community. I mustered the courage to open them. To my great relief, they praised our work. The Cuban Americans who wrote shared thoughtful, personal stories and the love of the island they had left behind. They understood we were doing science, work that would hopefully help keep their homeland healthy and beautiful as they remembered it.

Think Globally, Act Locally

It was Julio Baisre, in his former role as head of the Ministry of Fisheries, who pushed for a marine reserve at Gardens of the Queen. At the time, in the nineties, the area was heavily fished. He recognized the growing science about MPAs, suggesting that protecting an area from fishing could allow fish to recover and spill over to adjacent nonprotected areas where they could be fished. One such study examined Merritt Island National Wildlife Refuge, which encompasses the Kennedy Space Center in Florida. Because the area has been closed to public access and fishing for security reasons since 1962, it is, in essence, an "accidental" MPA. The study found that game fish were far more abundant inside the protected area than outside, and tagging studies showed that fish from inside the protected area traveled to the unprotected areas. Smart fishermen knew to drop their lines at the boundary line. Julio found a receptive audience at CITMA and discussions continued for several months. In the end, the plan moved forward. In 1996, 1,000 square kilometers of Gardens of the Queen were declared a marine reserve. In 2010, the area was upgraded to a national park and doubled in size to 2,000 square kilometers, the largest no-take marine reserve in the Caribbean. Only lobster fishing and catch-and-release fly-fishing are permitted. In addition to the local protections, most scientists agree that the fact that Gardens of the Queen is so far offshore—away from coastal impacts like pollution, sedimentation, and illegal fishing—is an important factor in its protection.

The result is considered a great success with strong recovery of the ecosystem and a thriving tourism business catering to scuba divers and fly fishermen. The divemasters, cooks, engineers, housekeepers, and crews do well financially and regularly stay aboard for a week before returning home. The government has partnered with an Italian tourism company to provide the services, but other than a few coordinators, the workers are all Cuban. The tourist operation is unfortunately a monopoly, with all the baggage that goes along with that. Fortunately, though, the company is supportive of Fabián and his important work.

In 2021, Fabián and his colleagues completed a comprehensive study that illustrates the success of the reserve. Gardens of the Queen has higher coral density and diversity and is in a comparatively healthier condition than in any of the other survey sites around Cuba. Fish populations are increasing, and their density and biomass are significantly higher than in other areas around the island. The study observes that Gardens of the Queen seems to be more resilient to coral bleaching than other areas around Cuba. Despite higher temperatures, the events are milder, and Fabián has pointed out that bleaching recovers within a few months.

I have quipped that if you want to see if MPAs work, stick your head below the surface of the Gardens of the Queen. Fabián's study would seem to bear that out. And it also appears that bigger is better.

So the logical question is, if you do right by reefs and manage as many of the local factors as you can, will that make them more resistant to global threats such as climate change? Nancy Knowlton believes so, but makes an important distinction: "I would say resilient rather than resistant. There's not a lot of evidence that suggests that protecting reefs from local stressors actually protects them from, say, a warming event that causes mass bleaching. In fact, we saw a good example of that at the Great Barrier Reef during the mass bleaching events in 2016 and 2017, even in a very remote section. So the Great Barrier Reef bleached severely with a lot of coral mortality. They were fine in terms of local stressors, but the water was so hot, it just cooked them essentially. But it is definitely the case that reefs that are protected locally are much better able to reproduce, and those small propagules, whether they're larvae or [coral] fragments, have a much better chance of surviving if they're not struggling to make their way against seaweeds growing all around them. So it's definitely the case that reefs are more resilient with protection."

So it would seem, following Nancy's logic and the results of Fabián's work and that of others, that the old adage, "Think globally, act locally," very much applies to coral reefs.

Tragedy of the Commons

Jim Bohnsack thinks about MPAs as an intersection of science, human behavior, and philosophy. Unaccustomed to a scientist venturing beyond the guardrails of science, my students listened intently as Jim spoke of a world of fish without protection: "The problem is, if I don't take, if I let this fish go, the probability is someone else is just going to catch it anyway. So why should I give it up? That's the 'tragedy of the commons.'" The term was originally coined in 1833 by British economist William Forster Lloyd who used the example of unregulated grazing on common land in Great Britain and Ireland. The result, he pointed out, would be people acting independently to serve their self-interest. Without regulations, such actions in the aggregate would quickly deplete the common resource, to the detriment of all (and the environment).

In this regard, Jim extols MPAs. They have become recognized as a "commons" that can benefit everyone. "People accept this on land. There's a flower bed out there in front of your building [at Johns Hopkins]. Why don't people pick them? Well, they know it's public property and the fact that we don't pick those flowers is accepted." Jim draws a distinction with the oceans. "We have a problem accepting that idea that we shouldn't fish [because it's also public property], and that's just hard for some people to grasp." Not all MPAs are "no take," that is, restricting all fishing. Instead, a range of restrictions are overlaid upon one another, the term of art being "ocean zoning." "It's a shared resource and lots of people use it for different things. Coral reef fish—people catch them and eat them, but also people spend time as tourists to go see them. That's a different use. And zoning is how we handle problems on land when there's conflicting uses or incompatible uses. And in an MPA it's the same in the water."

Jim also points to MPAs as being a safety net. "Having some areas protected provides an insurance policy, insurance that we won't catch them all; they'll be protected there. If we do screw up our management and have overfishing, we have something to rebuild the stock with. They're already out there. They can reproduce and correct our mistakes."

Paper Parks

For an MPA to work, it must be enforced, and that's been the Achilles' heel for many areas declared as protected areas. Without enforcement, they are protected on paper only. Cuba has declared 25 percent of its waters as protected areas, but my Cuban colleagues estimate only a small fraction of those are well managed and enforced.

Gardens of the Queen is considered a well-managed MPA. I had been told that there are regular patrols by the Cuban Coast Guard, but over the years, I had never seen a patrol boat. One day as I was returning from a dive, our captain, Arjelio, excitedly informed me that the Coast Guard boat was approaching on our starboard side. I envisioned a military cutter, perhaps with a camouflage paint job and a mounted gun or two. What I saw was a tiny barge-like boat, resembling a half-sunken Boston Whaler, with a small outboard and torn canopy. Two men in colorful shorts sat behind a small console at midships while a military officer in olive fatigues and a hat stood on the bow peering through binoculars. The boat was so slow it appeared that walking would be faster. As I pondered what I was seeing, it suddenly made sense to me. The tiny vessel in front of me did serve an enforcement purpose. The officer could make arrests, conduct inspections, write tickets, etc. But their job was impossible without the crews of the dive boats and the captains of the small, touristic fly-fishing skiffs whose eyes and ears were always watching and listening. They knew every boat that belonged in the reserve and had every incentive to report one that didn't. Their well-paying jobs depended on the health of the reserve. Friend and divemaster Andrés Jimenez would frequently make the point that a single boat could wipe out the reserve's shark population in a day. This "neighborhood watch"–type process directly addresses one of the challenges of enforcement: cost. It's an answer to one of Senator Whitehouse's chief concerns about protected areas. "For me, how you work out the economics of enforcement is a really, really big deal." Gardens of the Queen is an example of how providing economic incentives to those who depend on a healthy environment can help solve the enforcement problem.

Big Old Fat, Fecund Females

Jim Bohnsack offers another important reason for MPAs, this time challenging the age-old fisherman's adage to keep the big ones and throw back the little ones. Jim says we've gotten that completely backward. "Bigger is better," he observes. "In the marine environment, almost every time the larger you are, the more offspring you can produce." He offers a concrete example off the top of his head. "Let's just take a young 12-inch red snapper [one of the most popular commercial fish in the Gulf of Mexico and Caribbean—commonly found on both U.S. and Cuban plates]. They produce about 140,000 eggs. The probability though is all those are going to die, the chances of them surviving to adulthood to reproduce almost zero. You take a 10-year-old, 25-kilogram red snapper, and it's 60 centimeters, it's about twice as long but it's four times the weight. It produces 9.3 million eggs." He explains that one of these large fish, known as "Big Old Fat, Fecund Females" or BOFFFs, is worth 214 small fish. It's dramatic—let a fish grow larger and the eggs produced increases geometrically. "When fish are growing and they're small, a lot of that energy is for growth, to avoid predators. When they get old, they're too big for most predators to attack. So almost all their energy goes into eggs and production. So that explains why it's so important."

Yet we continue to target large, old fish—and dramatically so. As fishing fleets have decimated nearshore fish populations, they have ventured farther offshore and into deeper waters. Up to 5,000 feet below the surface, often inhabiting seamounts—underwater mountains reaching toward the surface—a fishery now exists for the orange roughy, a particularly long-lived and slow-growing fish. They don't reach sexual maturity until between 23 and 40 years, and even as large fish produce roughly 10 percent of the eggs of other fish. Orange roughies have been found as old as 250 years—born years before the U.S. Declaration of Independence was signed. And yet, these ancient sea dwellers, critical to the future of their species, end up in garlic butter and served to unwitting diners, unaware that the fish they're about to eat was swimming about when their many times great-grandparents were making their way in the world.

Many of the eggs laid by fish travel long distances, swept up in the currents. This is certainly true of a number of species, including snapper, spawned along the southern coast of Cuba. The strong prevailing currents sweep northward into the Gulf of Mexico, often forming a loop—the "loop current"—along the west Florida coast before turning east and following the Florida Keys, then turning northward as the Gulf Stream. I recall being aboard the *Suncoaster*, a research vessel of the University of South Florida, on an expedition where we were using submersibles to explore a deep reef 150 miles offshore. Early in the morning, sleepy and just waking up with a mug of coffee in hand, I made my way to the stern.

A large wake trailed behind us. I was confused. Were we under way? The engines were off. I looked forward and saw that we were anchored. I looked back at the wake of the ship and now comprehended just how powerful these currents are. When designing MPAs, it's important to consider things like currents that can create biological connections between distant places. Cuba and the United States are so connected, and modeling studies have shown that fish larvae can survive the ride to the States. That means Cuban fish could grow up to be American fish.

Jim points out that the importance of protecting BOFFFs doesn't end there. "Also, it's survived. Its genes have been successful. Generally, the larger females have more fat content and the eggs are better quality. It's gotten through all the things to be able to survive. So the problem for conservation is you have to have enough of those [BOFFFs] to survive to maintain the population . . . we're pulling the best genes from the sea and putting them in the ice chest."

Fresh Off the Boat

The fact that Cuba was able to declare 25 percent of its waters as protected areas came at the envy of many other nations. I've joked, "Cuba doesn't have any pesky public meetings to worry about. They can snap their fingers and the law is in place." Of course, the truth is that they do involve the public, but without strong independent private commercial interests in opposition and a strong environmental ethic among its citizens, there is little objection to such proposals. Cuba's 25 percent is among the highest worldwide, and until recently, dwarfed the percentage of U.S. waters in protected areas. That has changed in recent years, and, as mentioned many of these areas still lack enforcement and management plans, so they are protected in name only.

In the world outside Cuba, establishing MPAs can be a hockey match. I could think of no better example than the colorful stories of Buck Island National Monument on the East End of St. Croix in the U.S. Virgin Islands (USVI). As a professor I feel obligated to expose my students not only to the dry recitations of textbooks and scientific papers but also to the war stories of real people on the ground, their experiences, with the good, bad, and ugly that go along with them. So I recruited Nick Drayton and Stephanie Wear for the job.

I was fortunate enough to be Nick Drayton's supervisor at Ocean Conservancy when we established an office in the USVI. Originally from Barbados, Nick has a unique gift for building trust and friendship, even among professional adversaries. His humility and compassion are boundless, and with his gentle Caribbean accent, he offers thoughtful words of comfort and wisdom, all counterbalanced by a healthy and irreverent sense of humor. I consider Nick one of my closest friends and we've managed to maintain a long-distance connection over these many years. Through Nick, I got to know Dr. Stephanie Wear, senior scientist and strategy adviser at Nature Conservancy. I've admired Stephanie's work as she, for years, has dedicated herself to researching and developing strategies to protect coral reefs, and to understanding the complicated relationships between coral reefs and people. These days she's really into sewage, but we'll leave that topic for another day. She and Nick immersed themselves in the messy process of establishing the first MPA in the USVI beginning in 2001.

"I was *literally* fresh off the boat," Stephanie recalls. She was an intern for the first five months of her work in the USVI when she was thrown into the process of establishing the first-ever territorial marine park there. It began with her very first experience in a public meeting. "As Bill Clinton left office, he designated a whole bunch of national monuments. And one of those was Buck Island." The challenge was almost immediate. This was a popular fishing ground and Clinton's action was viewed by the community as a taking. "There was no consultation with the local community; it just happened. So there was tons of anger around the idea of protected areas, tons of suspicion." She recalls a lot of folding arms, radiating opposition throughout the room. "In a lot of Caribbean islands and small islands around the world, fishermen tend to have a very powerful voice and are very much listened to by politicians and leaders. It was all uphill; there was nothing that was making this easy on us."

Stephanie and Nick felt alone, and the promise of support from government agencies, university scientists, and other NGOs didn't amount to much. "I am such a huge outsider here. You know, I'm like 27 years old or something and totally ignorant to this whole thing. And I'm thinking, I understand coral reef systems. I've studied the science. I've worked with some of the best coral reef scientists in the world. I understand what this park should look like. I think I understand what we need to do. And I assumed that this was going to be some sort of scientific process. I walk in and I see this fully packed room. There were a couple of scientists, and the rest were pretty much fishermen." She recalls the government representatives giving a small, ineffective presentation, followed by passionate statements by angry fisherman after angry fisherman, expressing concern about their livelihoods, where they're going to fish, the traditions of their family and that they want their children and grandchildren to be able to fish.

Without warning, after promising not to do so, an official put Stephanie on the spot, calling on her to speak. "So I stood up, and I had no plan. I felt like I had a target on my forehead. And I said, 'We want your input, we want you to be a part of this. And it'll be a community-led process.' I probably said other things and rambled nervously." Immediately afterward, a large, imposing 60-year-old fisherman in overalls and a baseball cap with a big white beard and bloodshot eyes points directly at Stephanie. "And he says in front of all of those people, 'You are the DEVIL!' He scared the living daylights out of me, because he was quite a presence. And he was very much one of the outspoken people in that room." He continued, "This isn't a fair process, and we're not going to participate." That was when Stephanie came to an important realization that would shape much of her career moving forward. "This is not about science at all. This is about people. I listened and listened and listened, and that is one of the skills you need if you want to work in the conservation space. You don't walk in with your own agenda; you walk in to figure out what's going on with people and what their needs are and what their concerns are." And so, as the meeting ended, she resolved to do it. She walked up to her adversary in the parking lot to say, "Hi, remember me? I'm the devil." He chuckled and it opened a dialogue.

Nick shared his thoughts at that time. "I'm from Barbados, I'm a Caribbean person. How hard can this be?" He learned it could be very hard and that each island he worked on was unique. His take-home message, "Get to know your stakeholders. And help them to get to know you. Because if they don't see the sincerity and the honesty and the passion and the dedication and commitment from you, they'll smell it from a mile off." Their persistence—and humility—helped Stephanie and Nick forge close relations with other fishermen, so when new issues presented themselves, the foundation of trust was already there.

When he spoke to my class, Nick was managing director at Caribbean Creative Thinking, an organization that offers training in the use of

THE REMARKABLE REEFS OF CUBA

creativity in problem solving and innovation. "Being in conservation does require creativity. There's no manual. There's no playbook that we can go back to. And the goalposts keep getting moved. And so the thing that we need to be able to draw on is our inner creative skill. And that's what's going to get us through these challenging times." Nick then added, "It's never boring."

CHAPTER FIFTEEN

COLUMBUS AND THE INVISIBLE ISLAND

Even in winter an isolated patch of snow has a special quality.

—Andy Goldsworthy

The Invisible Island

A month after passing through the Gardens of the Queen, Colum-bus's ships were leaking, their provisions spoiling. It was clear that they would have to turn back. Anchored off Cuba's south-western coast near a large, mountainous pine-covered island, Columbus had seen enough. He was convinced Cuba was part of Asia and that return to Spain by land would be possible from the main Cuban island. He ordered each member of the crew to sign an affidavit testifying to this, and their signature bound them to have their tongue cut out should they ever contradict their signed statement. The next day, June 13, 1494, they landed on the nearby island. Columbus named it Evangelista. Over the centuries since, it has borne the names Isla de Cotorras (Isle of Parrots), Isla de Tesoros (Treasure Island), Isla de Pinos (Isle of Pines), and finally, Isla de la Juventud (Isle of Youth). While Cuba claimed its sovereignty from Spain in 1898, the fate of the Isle of Pines would not be settled until more than 25 years later when it officially became part of Cuba, though by then most of it was controlled by U.S. interests.

The 80,000 residents of the comma-shaped island often feel invisible, forgotten, and disconnected from the rest of Cuba, an "island within an

THE REMARKABLE REEFS OF CUBA

island" as they call it. Dwarfed by the massive main island of Cuba, the world is barely aware of its very existence, despite the fact that it is the seventh-largest island in the Caribbean, larger than St. Lucia, Barbados, Grenada, Bonaire, St. Croix, St. Thomas, St. John, Aruba, St. Barts, Saba, Terre-de-Haut, Isla Mujeres, and Key West combined. The island has struggled for decades, with limited economic opportunities for its residents, mostly in agriculture and fishing. It has long since been forgotten as an international tourist destination, when it was first advertised as one of Cuba's primary post-revolution tourist destinations in 1976.

The Cuban government has repeatedly tried to infuse life into the island. As Nick Miroff said starkly in the *Washington Post*, "Their island is a boneyard of big ideas." In 1978, Fidel Castro changed the name to Isle of Youth as part of an effort to bring new opportunity and meaning to the island. An initiative was launched to build a world-class international network of schools on the island, attracting students from Africa, Asia, and beyond. Thirty years later, in 2008, Hurricane Gustav, with sustained winds of 155 mph, decimated the island, laying waste to the international schools and that chapter of the island's development.

Yet, for all its struggles, the island is charming. There are no hotels—just private homes (*casas particulares*) for accommodations—and few restaurants. The residents are lovely and welcoming. The landscapes and seascapes are stunning, but visitors are few. The residents' feeling of isolation is palpable. A high-speed ferry to Havana and air service are available but the expense is out of reach for many.

Ministry versus Ministry

In 2015, Ocean Doctor led an expedition of Cuban and American scientists—including Jeremy Jackson—to visit the protected waters of the Isle of Youth; most of the southern half of the island is protected, part of Cuba's massive system of protected areas. As we dove among the coral reefs in the Punta Francés protected area, we indeed found some of Cuba's treasured coral reefs, gleaming and healthy. But we also found reefs in stress—some bleached white, some covered in slimy green algae, the telltale signs of a reef beginning to die. Fabián Pina's research bears that out—compared to Gardens of the Queen, algae cover is becoming much greater in that region. Our Cuban colleagues from CIM were surprised.

The situation had noticeably worsened since they had last been there just two to three years earlier. We saw no sharks, no groupers, and virtually no large predatory fish, a sure sign of the type of overfishing that contributes to a reef's decline.

Punta Francés has the same level of protection as Gardens of the Queen—no fishing is permitted. But here, there is virtually no enforcement. There is not even an enforcement boat. Commercial fishing vessels from Pinar del Río Province regularly fish within the protected area, something I have personally witnessed several times. In 2019, a group I was leading observed one such commercial fishing vessel, with expansive nets deployed across a narrow channel, pulling in hundreds of pounds of fish as they swam with the tide toward their feeding grounds, essentially swimming into the mouth of a funnel. It's common knowledge among our Cuban colleagues that this takes place and has for years. There's little doubt that such extensive fishing is a major factor in the degrading health of the reefs we witnessed. In many parts of the world, countries are struggling to deal with illegal, unreported, and unregulated (IUU) fishing. Cuba has been virtually free of illegal fishing in its waters by vessels from other countries. Ironically, it's their own fishing fleet that is breaking the law, a situation of Ministry versus Ministry: *Ministerio de la Industria Alimentaria* (MINAL), the Ministry of the Food Industry, versus CITMA.

As if living on Isle of Youth isn't isolating enough, a tiny community at the southern tip of the island stands isolated from the rest of the island's population. A journey that would take 20 minutes on improved road is a jostling three-hour journey across coral rock, mud, and broken pavement, the remains of the only road—unrepaired for 30 years—connecting the fishing community of *Cocodrilo* (Crocodile) to the rest of the island. In such isolation, opportunities are few for the community of 300 to 400 residents. Unemployment is nearly 40 percent. Economic need and a lack of alternatives encourages illegal fishing in the Punta Francés protected area, though the impact is certainly far less than that inflicted by the commercial fishing vessels.

Undersea Orgies

No sooner had I landed in St. Thomas, USVI, in 2004 than I was whisked away by Nick Drayton and fisherman Nicky Martínez to Nicky's awaiting skiff. It was rough. A constant fire hose of salt water blinded me, making it impossible to anticipate the oncoming swells. I tried to position myself in the sweet spot of avoiding being catapulted out of the boat or taking another painful blow to parts posterior. Our GPS readings were in agreement, but salt water caused a short in Nicky's GPS, and we found ourselves driving in circles. I marveled at how Nicky maintained his balance, driving with his tank on his back, his regulator and gauges swinging wildly. Getting dressed in scuba gear, reading GPS coordinates, and trying to keep lunch down was a challenge. Nick sighted the sun above the horizon and calculated 15 minutes to sunset. He shouted, "This is it . . . let's go!"

Nick's tank dislodged from his backpack before he could leave the boat. As he chased it across the bow, I told him I'd meet him in the water, where Nicky was already submerged and making his way to the anchor. I back-rolled off the side, but was hit broadside by an enormous swell. It ripped the tank from my backpack, but I continued down, intent to stay down even if I had to hand-carry my tank for the entire dive. Fortunately, Nicky quickly reattached it to my back. Then my dive computer malfunctioned. I set my watch and prepared to dive by the tables. At last, the three of us were below, on our way to 90 feet, but it soon became clear that the current was against us, and we were kicking furiously to keep on course. Still, the beauty and peace of the deep, dark waters at sunset were a welcome relief to the mayhem above.

Directly beneath me, a giant barracuda hovered motionless in the fading light. But he was not why we had come so far. We had journeyed to find a party . . . a party that only happens in the light of a full moon . . . a party with a singular purpose: sex. The inhabitants of this planet have found breathtakingly diverse approaches to perpetuating their respective species. Fish, like grouper, snapper, and triggerfish, among others, do it in groups—fantastic submarine orgies, forming spiraling columns from bottom to surface, releasing a massive cloud of eggs and sperm into the water. They are known as "spawning aggregations" or SPAGS for short, and they're in trouble. Fishermen have found them, and now GPS units have made it easier and easier to find them and return to them. In some ways, it's a fisherman's dream. All of the fish you could possibly want are concentrated below you. But at the same time, you're robbing from the future, killing the fish before they've had a chance to reproduce.

As a fisherman, Nicky is forward-looking. He is intent on protecting SPAGS and, in turn, protecting these fish populations for future generations—including his own children. We were there to map a suspected aggregation. We waited below as the waters darkened. Above us we suddenly saw triggerfish—10, then 20, then 30, spiraling and building into a column 40 feet above. But alas, they were only flirting. The group disbanded and we were left uncertain about whether we had found the right location.

As Nick and I sat out on the deck that night, shared a beer, and nursed our bruises while giving our instant replays of the events leading up to how we got them, a patch of clouds parted, revealing a brilliant full moon, casting its rays upon the glittering surf below. We both stared at the moon, without saying a word, knowing that somewhere out there, beneath the waves, a party was going on.

Back on the Isle of Youth, under the same full moon, fishermen have discovered SPAGS and are fishing them. Without enforcement, local fish populations could be decimated. Jorge Angulo is deeply concerned about this. He has spent years working in Punta Francés, as a teacher and a researcher. He knows the community well and they know and respect him. And Jorge is intent on protecting Punta Francés. He has observed fishing on the SPAGS, pointing out that the local currents and geography form a natural funnel. "Everything in the Gulf is just slowing down and aggregating in that little spot." It's ideal for fishermen. Having worked with Jorge for many years, I appreciate his direct approach. He knows the fishermen and engages with them, explaining the perils of fishing on SPAGS, and telling them that the fish will be gone within 10 years. So he's proceeding incrementally, asking them, "Why don't you just release at least 30 percent of those fish full of eggs ready to spawn?" He hopes that they will gain a better understanding—and respect—for the fish they are taking.

How Much Is That Coral in the Window?

By the 1960s it became clear that traditional economics failed to take into account important factors, such as social welfare and the environment. For years, we considered our natural environment as "free." We dumped our toxins into rivers and streams and extracted minerals and wildlife from the land

and oceans without considering the cost of our actions. Environmental economics seeks to understand the economic value that the natural environment provides, helping to address the shortfalls of policies based on traditional economics that place little or no value on the health of natural ecosystems. Had we placed an economic value on keeping our natural ecosystems intact, there is no doubt that many decisions would have been made much differently. For example, had we adequately considered the value of the Everglades to the economy of South Florida, we might not be engaged today in the largest environmental restoration project in history to try to save them.

Interestingly, Cuba's "Law of the Environment" requires that its environmental ministry (CITMA) "direct actions intended to promote the economic evaluation of biological diversity." Cuba has a handful of dedicated environmental economists who have had to adapt to an ever-shifting economic landscape that is comprised of dizzying combinations of socialist and capitalist elements. Tamara Figueredo's use of environmental economics helped support the Cuban government's decision to establish Gardens of the Queen as a national park in 2010.

Applying the principles of environmental economics could serve to "future-proof" the country's strong environmental legacy against future economic pressures by providing it with the tools and information necessary to demonstrate the economic value of its resources in their natural state. Such studies, such as those in Gardens of the Queen, have already demonstrated that protecting large marine areas can be more valuable to the economy than commercial fishing, thanks to Cuba's growing ecotourism sector. Studies around the world show similar results. A study in Palau shows that over its lifetime, the value of a shark to tourism is $1 to $2 million, versus a few hundred dollars carved up and tossed on dinner plates. World Resources Institute found that in Belize the economic contribution of mangroves and coral reefs for shoreline protection is nearly $350 million annually. A 2014 study shows that coral reefs reduce wave energy by an average of 97 percent. The same study reports that the cost of building artificial coastal barriers is 15 times higher per square meter than coral reef restoration. It's not hard to see the policy implications—protecting coral reefs not only benefits humans; it's a no-brainer business decision, especially in an era of sea level rise and monster hurricanes. And that's before considering the value of coral reefs and mangroves for tourism and fishing.

Taking a cue from Gardens of the Queen, where workers and the local communities from which they come have financial incentive to protect their waters, could the same incentives be created for Cocodrilo to protect Punta Francés? In 2015, we began supporting the community of Cocodrilo, exploring new ideas that could both provide new economic opportunities while also protecting their coral reef ecosystems. From Cocodrilo's residents themselves came an idea they called Project "Red Alerta." As it happens, it's a bit of a play on words, the literal English translation is "Alert Network," but also conveys the urgency of protecting coral reefs with its similarity to the phrase, "red alert." The brainchild of Red Alerta is community leader and scientist Reinaldo "Nene" Borrego Hernandez. He is admired on the island and by his colleagues at CIM. Unassuming, soft spoken, and kind, he is dedicated to helping his community flourish, and he has great ideas that use science as a catalyst to bring economic benefit to his community. The idea of Red Alerta is to train community participants—primarily young people—to snorkel and scuba dive, to identify corals and fish, and to monitor their coral reefs, providing important monitoring data to scientists back at CIM. The data will also be incorporated into a regional coral monitoring network to help scientists fill an important gap in our understanding of the health of coral reefs throughout the Caribbean. Until recently, data from Cuba has been missing from such regional monitoring efforts. Then, using the same skills, Red Alerta participants would be able to host visitors as guides for ecotourism, an activity that would support their local economy while providing strong incentive to protect their reefs. The project would also bring exchanges and knowledge from other communities in the Caribbean and Latin America, sharing ideas and experiences related to sustainable tourism.

Nene and his business partner have already established the first *casa particular*/dive center in Cocodrilo and I was honored to be the first American to stay there. It's a charming, modest cinder block building with two bedrooms. "Dive center" is a bit of a stretch, but they have all the basics to take small groups snorkeling or scuba diving. Though I did lose some blood, I slept peacefully beneath vintage 1980s regulators hanging from their hoses above me. I'm happy to say they have since installed mosquito netting on the windows.

Small-scale ecotourism, with the charm of a place like Cocodrilo and other parts of the Isle of Youth, could provide the economic incentives and capacity for the island's communities to protect their natural areas. Cocodrilo is already doing it. Ocean conservation is now part of the curriculum at the small school there. We helped support the construction of signage with slogans like, "We protect our beaches and coastal waters." Nene leads regular beach cleanups—above and below the water—that the youth relish.

Jorge Angulo has given a lot of thought to this, focusing on the areas outside Cocodrilo. "I would say the first thing would be to try to recover the [Colony] Hotel and the Marina." The small hotel is practically empty but sits on an idyllic beach, just down the road from an underused dive center. During the summers, Jorge's students stay at the hotel, so, like Nene, he sees the value of a nexus between science and tourism, that the two can coexist and help revive the local economy.

Ellen Rugh, director of research and programs at the Center for Responsible Travel in Washington, DC, offers additional advice with an eye toward achieving sustainable, responsible tourism. "Just because you're on a bird-watching tour, just because you're kayaking or doing something in nature, doesn't mean that the business operations themselves are fully sustainable." She posits that it is important that ecotourism operations work "to preserve the nature around them, to protect their local communities and uplift their local communities."

From Columbus to Castro

From Columbus to Castro and all the chapters in between that have tried to define this island of so many names, Isle of Youth remains a paradox: It is an island that is enormous yet virtually invisible; an island nestled close to Cuba's mainland's coast, yet exists in virtual isolation; an island with a history of grand dreams but strewn with heartbreaking failures; an island of profound natural beauty but, despite its isolation, now faces some of the same ecological threats the well-known, oft-visited islands of the Caribbean face. Now Isle of Youth enters another new chapter. This time many of its hopes lie in a tiny, charming community, distantly isolated from the rest of us, whose bright-eyed residents desperately want to make a difference. With vision, hope, and hard work, we hope to help make their dreams—and the dreams of this beautiful island—a reality at last.

PART FIVE

SEÑOR, THE WORLD IS ABOUT TO CHANGE

At its best, life is completely unpredictable.

—Christopher Walken

CHAPTER SIXTEEN

TAKING IT TO THE STREETS

Nothing is certain in Cuba until it is in the past.

—Nino, Cuban boat captain

A New Era

The morning of December 17, 2014, was typical. I stepped into the shower at my friend's apartment near the famous Tropicana nightclub to a gurgling sound and barely three drops of water. So, in typical Cuban fashion, we walked several blocks to the home of our friend's parents who were gracious enough—actually delighted—to let us use their shower.

Clean and dressed, I left for my meetings. Normally, I would take an *almendrone*, literally "almond tree," the name Cubans have affectionately given the 70-year-old American cars due to their almond-like shape. Most are repowered with diesel engines and used as taxis, at times cramming seven, eight, or nine people in for an un-air-conditioned sojourn through Havana. A strange marriage of the ancient analog world and the modern digital world, many *almendrones* are equipped with mood lighting and modern, high-quality stereo systems booming reggaeton and making conversation practically impossible. I began using *almendrones* regularly to save money, as those of us working for nonprofit organizations strive to do. A drive across town in a conventional taxi could cost $15; take an *almendrone* and it would set you back 50 cents. Each *almendrone* has a specific route but no markings to indicate which. To hail a car on your desired route requires an elaborate set of hand signals, first signaling the route and then the number of passengers. To be sure, you then poke your head in the passenger window and ask the driver, "*Hasta 66?* (Do you go as far as 66th Avenue?)"

That morning I was attending the second day of an annual conference held by El Instituto Superior de Relaciones Internacionales, the Higher Institute for International Relations (ISRI), and Centro de Investigaciones de Política Internacional, the Center for Research of International Politics (CIPI). With increased recognition for their role in improving Cuba–U.S. relations, a number of U.S. NGOs were invited to speak about our work, including marine science. The room was full of the world's leading experts on U.S.–Cuba relations, along with high-level representatives from the Cuban Ministry of Foreign Relations.

In a hurry after my side trip to get a shower, I opted to hail any private car that might pick me up, a common though illegal practice among Cubans. A Russian Lada pulled over. I hopped into the front seat and negotiated a price of four Cuban Convertible Pesos (a little over four dollars) for the ride to the conference. As we pulled up and I opened the door, the driver told me, "Señor, the world is about to change today." I looked at him quizzically and could tell he wanted to leave me in suspense.

I entered the conference hall, but instead of the first panel of speakers assemblying there was commotion in the front of the room, a buzz filled the air, pregnant with an announcement of something significant. But even the experts from leading U.S. universities who had pontificated on Cuba–U.S. relations and offered their reasoned predictions the day prior were completely in the dark. A large, flat panel TV was rolled to the front of the room, the lights dimmed, and the image of Cuban president Raúl Castro sitting behind his desk appeared. He announced the beginning of a new era of relations between the United States and Cuba. Normalization of diplomatic relations between our two countries would commence. The so-called Cuban Five, imprisoned in the States as suspected Cuban spies, had been repatriated, having landed in Havana earlier that morning, just as Alan Gross, imprisoned in Cuba for illegally installing electronic equipment allowing Cubans internet access, landed in Miami. Immediately following Castro's speech, one of the organizers inserted a USB drive into the TV and began to play a speech by President Barack Obama, which had been aired simultaneously.

As the dumbfounded experts sat in disbelief, shaking their heads, an ISRI student in the back of the room stood, fist in the air, and shouted political slogans, cheering this historic moment. She was quickly joined by dozens of other students, chanting political slogans. One of the students grabbed the ISRI banner and suddenly a river of students swept down the stairway and into the streets. A number of us Americans joined in as the students continued chanting, the ISRI banner proudly displayed in front of the procession as it made its way throughout Havana with the students. American flags waved from balconies. Seniors watching from the sidewalk wept. Others cheered. A cacophony of shouting, music, and blowing horns from 1948 Buicks, 1987 Moskvitches, and 2010 Elantras filled the air. It was an intoxicating moment—figuratively, and for many, quite literally.

The next day, a revered figure attending the meeting, Wayne Smith, the former head of the mission for the United States in Havana under Carter and Reagan (and who ultimately resigned in protest to U.S. policy toward Cuba), joyfully faced the audience and said, "I guess they finally heard me!"

CHAPTER SEVENTEEN

THE HANGOVER

Yet nothing resembling the near-complete destruction of previous societies in the Caribbean and their replacement by new and unprecedented political, economic, and cultural orders had ever been envisioned or achieved elsewhere before.

—Stephan Palmié and Francisco A. Scarano,
The Caribbean: A History of the Region and Its Peoples

Cuba *Autentica*

When a foreigner sets foot in Cuba, it immediately becomes clear that this magical island is profoundly unique and has developed much differently than any other country in Latin America and the Caribbean. And for those who venture into its verdant mountains or below its aquamarine waves, a striking revelation awaits: Just as the fifties-era Chevys and horse-drawn buggies portray an island seemingly frozen in time, so, too, do its exceptionally healthy and vibrant ecosystems illustrate that Cuba may have picked the perfect time in history not to follow the path of its neighbors, especially the path of mass tourism.

A few months after the dramatic announcement of restoring diplomatic relations, I found myself a few blocks from my DC home in an enormous crowd, enduring oppressive heat and humidity typical of a July day in Washington, DC. With each tug of the rope by Cuban foreign minister Bruno Rodriguez, the Cuban flag inched upward, finding a slight breeze, and proudly showed off its brilliant colors of red, white, and blue to the 500 or so onlookers. The Cubans and Cuban Americans—never known for their silence at public events—beamed with national pride and shouted with joy as the flag crept up, "Fidel, Fidel!!" Countless eyes filled with tears. Many embraced. The world was changing before us. The Cuban flag flew in Washington, DC, for the first time in 54 years, signaling the reopening of the Cuban embassy and the beginning of normalization of relations with the United States.

Inside the embassy at the reception that followed, we hoisted mojitos and exchanged congratulations. But a number of us had long anticipated this moment with both joy and worry, realizing that the United States could become a greater threat to Cuba as its friend than it ever was as its enemy. Many of us have heard a common refrain from acquaintances, "I want to get to Cuba before the Americans ruin it." Indeed, there is a great fear that Cuba could end up like Cancún and many other places in the Caribbean that have destroyed their coral reefs and lost their culture and identity in the process. By not developing like the rest of the Caribbean, Cuba has so far spared its natural ecosystems, including its coral reefs. However, a flood of tourism and business development from the States could undermine Cuba's natural heritage and culture. In the conservation community, our euphoria about the new era of relations between our countries was quickly supplanted by a painful hangover in anticipation of the possible catastrophic consequences of the *tsunami Norteamericano*, as the Cubans put it.

Among many reforms, President Obama reinstituted a legal category of travel created by the Clinton administration known as "People to People Travel," which required "meaningful interaction" between Americans and the Cuban people. Although the stated goal of this category was to "promote independence" from the Cuban government, such travel was welcomed by Cuba as it brought in millions of dollars for the Cuban economy. Ocean Doctor took advantage of this by bringing hundreds of travelers to Cuba to see firsthand the island's healthy coral reefs while enjoying "meaningful interaction" with Cuba's scientists, ecotourism professionals, and others. By 2017, U.S. visitation would be 10 times what it had been in 2010. Ironically, many of the tourists during the post-Obama and pre-Trump period were from non-U.S. countries, seeking to visit Cuba "before the Americans ruin it."

How the Hell Do You Translate "Spring Break"?

In 2018, Ocean Doctor released a report, "A Century of Unsustainable Tourism in the Caribbean: Lessons Learned and Opportunities for Cuba." The report was rolled out with great fanfare in an event including Cuban ambassador José Cabañas, the Center for International Policy, and attorney Robert Muse. In consultation with our Cuban colleagues, we sought to shed light on Cuba's unique opportunity to leapfrog the mistakes of mass tourism made by other Caribbean nations that have profoundly reshaped the Caribbean. According to IUCN, landscape modification due to tourism development is one of the main contemporary drivers of habitat loss in the Caribbean. There has been a 42 percent loss of the region's mangroves in the past 25 years. The Caribbean is now the most tourism-intensive and tourism-dependent region in the world.

Caribbean tourism has its roots in the 1800s when resort hotels appeared in the Bahamas, Jamaica, and Barbados. By the late 1800s dual-purpose banana boats were used to deliver tourists to the Caribbean. Until the mid-20th century, Caribbean tourism remained modest until, in response to the decline of dominant plantation economy, largely sugarcane, the United Nations and World Bank, beginning with Puerto Rico, the Bahamas, and Jamaica, began to invest in tourism as a tool for development and promotion of robust economies. On its heels was the emergence of the economical long-haul jet airplane, which, for the first time, put the Caribbean in reach financially for the average vacationer. Visitors from Europe could now make the journey in eight hours rather than weeks by ship. For many Americans, just a couple of hours in the air and you'd be sipping an ice-cold piña colada in the tropical sun. From 1961 to 1971, visitation increased sixfold to nearly 5 million visitors per year. But by the 1980s, the distinctions among the different countries of the Caribbean began to blur as intense competition among the islands ensued. Soon, advertising offered "sun, sand, and sea" tourism. Travel to the Caribbean became commoditized, the distinctions among islands hazy, and consumers more interested in a competitive price and less interested in which country the beach they'd visit was attached to.

Today, high-volume mass tourism dominates the Caribbean tourism market with prepaid travel packages for resorts and cruise ship vacations. Culturally, the region has experienced homogenization—local cultures and traditions supplanted by what locals feel the tourists want to see. Some islands have even moved the dates of holidays to correspond to tourist visitation.

The region has the highest level of "leakage"—the loss of tourism revenue to other countries—than any other part of the world. Of each tourist dollar spent in the Caribbean, only 20 cents stay there. Offshore multinational corporations take the rest. Such corporations commonly import their workforces, leaving low-wage and unskilled jobs for the locals. This was the case when I visited Tobago during a troubling period when young people were slaughtering leatherback sea turtles. It was their way of protesting a new Hilton that appeared on what they considered the best beach on the island. The hotel came with reassurances of economic opportunities, jobs, and other benefits for the island. Instead, locals were no longer permitted to visit that beach, the good jobs went to nonnative corporate employees, and, to make matters worse, the hotel was all-inclusive, meaning that guests had little incentive to visit restaurants and other businesses in town. But why slaughter sea turtles? They explained that it was so the tourists couldn't enjoy them.

Cancún, one of the Caribbean's most popular destinations, is a vivid example—perhaps the poster child—of the consequences of modern-day mass tourism. A successful agriculture economy has been displaced along with its community and culture. The beach is lined with multinational hotels. Its reefs are dying. And then there are the spring breakers. During a presentation in Havana, as best I tried, I could not convey the concept of spring break in Spanish or any other language. My audience sat silent with polite confusion. Finally, I downloaded an image of hundreds of scantily clad drunk teens wreaking havoc on a Cancún beach as an example of what the city has become. I keep the photo in my presentation because I still find it impossible to translate "spring break."

Opportunities for Cuba

Prior to the revolution, Cuba was a popular tourist destination, especially for Americans. To my surprise, my grandmother showed me a photo of her on a beach in Havana in 1930—she was on her honeymoon. Later, Cuba became host to an era of organized crime from the United States. Following the revolution, the country moved away from tourism, albeit with a number of exceptions, especially for its Russian visitors. However, during the devastation of the Special Period, with few options available, Cuba hitched its economic wagon firmly to tourism and saw it become a dominant part of the Cuban economy, today second only to the country's export of professional medical services. Although Americans are largely absent, Canadians, Italians, Russians, and Germans are among the island's international visitors.

While Cuba has developed large resorts like other Caribbean countries, they are localized in areas such as Varadero, a small peninsula lined with large hotels and masses of tourists on the beach, an area reminiscent of Cancún. The original community is long gone, and any hint of Cuban culture seems staged and stereotyped. Thankfully, the vast majority of Cuba's coastline is resort-free. Cuba has not been culturally homogenized as has the rest of the Caribbean, and its natural environment thrives. The Cuban Ministry of Tourism's slogan is *"Autentica Cuba"* (Authentic Cuba). There's no need to go the path of Cancún and remake the Cuban landscape and communities to serve tourism. Cuba need not become a sun, sand, and sea commodity. What's special about Cuba is its unapologetic authenticity, and travelers are willing to pay a premium for a truly authentic experience—a healthy, vibrant, natural environment and rich culture.

My Cuban colleagues have pointed out that Cuba has strong environmental laws, strong foreign investment laws, and has been open to the rest of the world for many years. Surely, they are ready for the Americans. There's truth in that position, and Cuba deserves praise for its strong, science-based environmental laws. However, the onslaught of millions of American tourists and the promise of billions in foreign investment will surely create unprecedented pressures. Before the Trump administration's tightening restrictions on Cuba travel, followed by the pandemic, it felt to many of us—Cuban and American—like a footrace to chart a sustainable course for the wave of tourism that Cuba faced and no doubt will again face.

One of Cuba's unique opportunities is to develop small-scale tourism opportunities rather than follow the mass tourism model. Due to changes in government regulations over the years, the rental of private homes, *casas particulares*, has become permitted and popular. With privacy, creature comforts, a personal touch, and tranquility not possible in large hotels, my travelers overwhelmingly prefer to stay in a *casa particular*, getting to know the owners and the neighborhood, and being doted upon and spoiled with fresh mango, guava, and other delicacies. For the owners, it's a good income and has the benefit of keeping much of the income local, so the entire community benefits. But Cuba is at an economic crossroads, with the danger of succumbing to the false promise of mass tourism.

A close friend and exceptional Cuban tour guide, Jesús Noguera, is distressed to see a wave of large hotels being built in Havana. He aspires to be an independent tour guide catering to small groups and offering a more personal experience. But, despite a trend in Cuba to permit more *cuentapropistas*, private entrepreneurs, "tour guide" is not on the list of permitted activities. "It's not reflecting the reality of our time. In Cuba, there is Airbnb, there is internet, there is now more freedom. I want to look for a client and offer my services as a tour guide for five or six people." He compares such an experience with a state-owned bus packed with a group of 40 that has run out of gas. Commenting on my two-year absence from Cuba during the pandemic, Jesús warns, "When you come back you will realize how many new hotels are being built . . . ugly! Not a single Cuban architect was hired to design a beautiful facade. It's a piece of concrete and glass facing the Atlantic Ocean with not a single balcony." Jesús recently texted me a photo of his new prize possession, what he calls "The Mimi," a 1948 Dodge Custom, a full-sized four-door sedan, which holds seven passengers. He hopes to use it for small tourist groups.

In its 10-year plan, Cuba is considering what would amount to the highest number of all-inclusive resorts in the Caribbean. Fortunately, important research is under way at the University of Havana's Department of Tourism to look at alternatives. Under the direction of the associate dean of Research and Postgraduate Studies and associate professor and researcher Dr. Lisandra Torres Hechavarría, small-scale models of tourism are being examined, including those where several communities can cooperate with one another and work together to serve the needs of tourists. This could even include agricultural linkages, where one community exports food to feed tourists staying in the neighboring community. Such an approach strikes a balance between tourism and preserving the character and culture of local communities. Ellen Rugh at the Center for Responsible Travel underscores the importance of such solutions: "They minimize the negative social and environmental impacts. And it really just helps local people conserve fragile cultures."

PART SIX
LAND OF HOPE AND DREAMS

Hope smiles from the threshold of the year to come, whispering, "It will be happier."

—Alfred, Lord Tennyson

CHAPTER EIGHTEEN

HOPE AMID MAYHEM

The work goes on, the cause endures, the hope still lives, and the dreams shall never die.

—Edward Kennedy

Master of Disaster

The envelope please. And the winner of the 1953 Oscar for best documentary is . . . *The Sea Around Us*. Accepting the award was the film's director and writer, Irwin Allen, best known for disaster films like *The Towering Inferno* and the *The Poseidon Adventure*. This was a screen adaption of Rachel Carson's book of the same name. She is most well known for her 1962 landmark book, *Silent Spring*, mentioned earlier, her exposé documenting the environmental impacts of pesticides and the evils of the chemical industry. Her earlier book, *The Sea Around Us*, presented a survey of what was known about the oceans, inspiring curiosity and the wonder of discovery.

When I finally got my hands on the film version, I was horrified. One of the first scenes is of a skin diver bearing a long spear, chasing a manta ray. The music turns menacing as the narrator describes the scene. "The aqua lung, the snorkel, the underwater spear, and swim flippers have in combination given man the agility, the breathing equipment, and the striking power of even the fiercest of the underwater monsters. Here the great manta ray appears to be in a playful mood. This should not be

203

mistaken for friendliness, for the manta when aroused is among the more dangerous of all the undersea game. Armed with his underwater spear, a man pursues the giant. This is one of the most dangerous of sports. Unless the man's first shot is true, he runs the risk of a watery death." Manta rays are, for the most part, harmless to humans. The drama then continues: "Equally as deadly and twice as cunning as the manta is the vicious moray eel." A moray eel is shown struggling, having been speared by a diver. "If in all the seas there's a creature hated more than the shark by fish and fishermen alike, it's the moray eel. All the fury of the sea explodes in this snake-like killer. The jaws snap, the tail thrashes, the hate becomes a living thing!" Moray eels are a favorite of divers and bite only if disturbed. "New danger, a killer shark is attracted by the blood. The man attempts to escape . . . look out!" The skin diver is seen tearing open an obviously already deceased shark's belly with a large dive knife as blood pours into the water. "The fight is short and one sided." The documentary continues, showing the "brave" whalers with explosive harpoons mounted aboard their ship, pursuing a whale. If I'm showing the film clip in public, I make it a point to stop the film before the harpoon fires.

As disturbed I was to see this shit show of a documentary, it ultimately left me content, realizing how far we have come and how much our attitudes about the sea and its inhabitants have changed since 1953. Dive magazines run ads with divers photographing, not spearing, morays, and more people than ever know that sharks have more to fear from us than we do from them. For years it was next to impossible to get media coverage about the oceans. Today there is more awareness than ever before, and if we are vigilant, their future doesn't have to be fodder for yet another disaster film. Saving coral reefs begins with awareness, and we've made progress. Climate change has gone from an obscure, ridiculed "theory" to an international crisis that has summoned world action. And those leading the charge are the "brave" ones.

The Arrogance of the Present

Despite tremendous progress in our attitudes about the sea, some outdated attitudes about the oceans persist. The disbelief that fish and other sea creatures could go extinct stems largely from early perceptions of the ocean. Antiquated myths about ocean life's immunity to extinction form some of the basic assumptions underlying today's ocean resource management programs and public policy. In 1884, Thomas Henry Huxley commented that probably all of the great sea fisheries were inexhaustible and "that nothing we do seriously affects the number of fish. And any attempt to regulate these fisheries seems consequently . . . to be useless." Today, we chuckle at the absurdity of Huxley's comment as fish populations are seriously overfished around the world. But none of us was alive during Huxley's time. And like Columbus's observation of sea turtles, pulled from his logs by Jeremy Jackson, we must face the fact that we have never experienced what undisturbed oceans looked like. We have no reliable baseline, and thus we use the baselines from our own lives, a phenomenon known as "shifting baselines," which Jeremy sometimes refers to acerbically as "the arrogance of the present."

On a dive with Nick Drayton in 2003 near Conga, a tiny island near St. John in the USVI, we surfaced and climbed back into the skiff. We looked at each other with disgust. The reef looked positively awful. As we stowed our gear, a mammoth tourist boat, *Wild Thing*, approached and anchored nearby, dumping 50 snorkelers into the water. They shrieked with joy through their snorkels, occasionally surfacing to exclaim how awesome the reef was. They had probably never seen a coral reef. This was their baseline. And it sure beat the hell out of the snowdrifts that awaited them back in Minnesota.

Jeremy is passionate about the topic. "If you were to tell an ecologist that on the prairie buffalo were never abundant, they'd laugh at you. They'd say, sure, we don't have quantitative data from the 15th century, but we know that people described herds that took all day to pass. And we know how many skulls are out there lying on the prairie. But marine ecologists were doing the equivalent of asserting that there never were abundant buffalo in North America. And that's the shifting baseline syndrome: that everybody thinks natural is the way it was when they were a kid. Unnatural is the way it is when they're old, which is why older people are more depressing than younger people. And then because you're young people, you don't listen to your parents, you make the same mistake, you screw up in the same way. And generation by generation, we totally forget the way things used to be."

Restoring the Everglades is a challenge of shifting baselines, too. What will it be restored to? Shannon Estenoz observes, "As we restore it, people will say, 'I spent my whole life in the Everglades and this is what it's supposed to look like.' And we always say, 'Well, unless you're 120 years old, this isn't really what it's about. You don't have a living memory of what the Everglades is really supposed to look like.' "

Thankfully, and thanks in large part to Jeremy, there is growing awareness of shifting baselines. Senator Whitehouse sees the potential of the shifting baselines syndrome to help people set their baselines high. "Once you've seen a truly vibrant and healthy reef, you now have a new standard to compare it to. I remember on one of my Florida climate trips meeting a mayor [in the Florida Keys]. And I asked her, 'So how are your reefs doing?' And she said, 'Well, they're beautiful . . . unless you were here 10 years ago.' So when you see a place where it can be preserved, it resets your expectations. It's unforgettable. And it sets the bar for what can be done. Part of the problem with the ecological and environmental degradation of the world is that you begin to lose sight of what it was, you keep readjusting to the new low, you keep dumbing down and at some point it becomes very dangerous. And that's where seeing clearly what a proper healthy, vibrant living reef looks like is so important." I'm glad I was able to show Senator Whitehouse the healthy reefs of Cuba. No doubt it's helped push his expectations even higher. When I received the call from *60 Minutes* associate producer Anya Bourg, she wanted her team to join us in Cuba to tell the story of coral reefs. Otherwise, she insightfully reasoned, the public wouldn't know what a healthy reef was supposed to look like. In so doing, she helped set the baseline high for 19 million viewers.

More and more, the term "shifting baselines" is found in the scientific literature, no longer ignored as in the past. And new tools, like Jeff Orlowski's time-lapse imagery, are helping the public see changes unfold and perceive how rapidly our baselines can change.

Cuba's Living Lifeboat

I've described Cuba as a living laboratory. The reasons for its healthy reefs fall in line with what we know is killing reefs elsewhere. What's shocking is just how healthy and beautiful many of Cuba's reefs are. Cuba has turned, by necessity, to organic farming, largely avoiding the devastation of nutrient pollution and the fueling of uncontrolled algae growth. It has thus far avoided the perils of mass tourism and its appetite to chew up and spit out coastal ecosystems. It has established the largest no-take MPA in the Caribbean and has declared 25 percent of its waters as MPAs. Parrotfish are safer in Cuba—especially within its MPAs. Cuba has virtually phased out bottom trawling. The country's fishing is largely selective from small boats, though there is still a ways to go before fishing is sustainable and nonthreatening for coral reefs. Cuba's reefs have experienced low levels of disease, perhaps because many are farther from shore and the fact that there is less coastal development. The health of the reefs may also play a role in their resistance to disease. Cuban environmental laws are strong, and the legal mandate to use environmental economics in environmental decision making can help find environmentally and economically sustainable alternatives for those in coastal communities, alternatives that provide incentive to protect—not exploit—their coral reef ecosystems. *Diadema* are recovering in Gardens of the Queen—slowly, but faster than most areas in the Caribbean. Healthy marine ecosystems with healthy predators may prove to be the best defense against invaders like the lionfish. Taken together, many find it hard to deny that Cuba's coral reefs are more resilient than those of their counterparts elsewhere in the Caribbean, and resilient to a warming climate . . . for now.

Cuba is not only recognized as a living laboratory. It is now recognized as a living lifeboat as corals around the world continue to die. Twenty-first-century coral conservation is turning to triage, recognizing that we simply can't save all of the coral reefs in time. "The situation with coral reefs is probably the most challenging of all the marine ecosystems on the planet," says Nancy Knowlton, and she asks, "Where should an NGO focus its efforts if it's trying to make a difference for reefs, in places where there's nowhere the financial resources aren't available? Now, there have been some fairly sophisticated modeling efforts aimed at figuring this out." Nancy is referring to a 2018 study led by Hawthorne L. Beyer of the Global Change Institute at the University of Queensland that identified coral reef locations globally that are likely to have an increased chance of surviving projected climate changes in the absence of other impacts. The team identified 50 reefs worldwide, which they termed "coral refugia." Fully 10 percent—five of the 50—are located in Cuba, including Gardens of the Queen. The reefs identified include those with the highest level of health, adaptability, resilience, and connectivity to other coral reef ecosystems. Such areas can help repopulate degraded areas downstream. But it's a race, all dependent on stabilizing climate change in time. It's one part of the solution. More 21st-century science and solutions also offer hope.

Adapting to Hot Water

Corals can naturally adapt to a warming climate. Nancy Knowlton has observed interesting transitions along thermal gradients from Tokyo to Florida. Such adaptation offers hope, but it tends to be slow and uncertain, and we may not have the luxury of waiting and wondering.

Nancy told my students about cutting-edge science that uses selective breeding to produce strains that are disease-resistant or thermally resistant. "We're getting a much better understanding now of the genetics of resistance to warm temperatures." Some combinations of microalgae in coral tissues can make corals quite resistant to warm temperatures, perhaps as much as a 2-centigrade temperature increase. "There are programs around the world seeking to breed these corals that might be more resistant to future conditions. There's not really a single silver bullet with regard to this, but to the extent that these corals can be essentially genetically engineered or selectively bred to be more resistant, then they can be used to propagate corals. You can't do this at a scale to replace the entire Great Barrier Reef, and some people, for that reason, have been very skeptical about these efforts. But that's not really the point. What you want to do is get pockets of reef being resistant, and then on their own over time they can recover." Nancy smiles—revealing her ever-present optimism, telling the students that she finds these new approaches encouraging. "The genetic toolbox now is just unlike anything that was available when I was a grad student, or even 10 years ago. For example, they're now using CRISPR technology [a tool to edit genomes and alter DNA sequences] on corals to figure out which genes are most responsible for susceptibility or resistance to high temperatures."

The Whale Whisperer

The doors opened, and the elevator at the International Monetary Fund (IMF) headquarters in Washington, DC, was packed. Dr. Ralph Chami, assistant director of the Western Hemisphere Division, Institute for Capacity Development, managed to squeeze aboard. One of the IMF staff, smirk on his face, welcomed the "whale whisperer" aboard, clearly intending his words to land pejoratively. Ralph recalls, "He said, in earshot of the fellow passengers, 'So when do they whisper to you?' I said, 'Every day!' He said, 'Really? Well, what do they whisper to you about?' " Ralph gives me a devilish smile as he recalls every word of the exchange. "I said, 'Would you like to know? Come closer, because it's a whisper. Come closer.' So he actually does this, and I whispered, 'He whispered to me that you're a jerk!' I whispered it out loud. The whole elevator exploded!"

Ralph's epithet came innocently enough—as a high-level, 23-year IMF employee working on, of all things, whales, which had many of his colleagues scratching their heads. For some, it manifested as ridicule, like for his colleague on the elevator. Such insults are often the case for pioneers. Far from a strategically planned direction for the IMF, Ralph's work was born accidentally aboard a skiff in Mexico's Sea of Cortez, his guide, Michael Fishbach, former professional tennis player and now executive director of the Great Whale Conservancy where I served as board chair. So I suppose it was the blue whale that introduced the two of us.

Ralph emotionally described his close-up encounter with these magnificent animals: "The largest creature that ever lived is feeding next to you, and breathing. That breath! It's wondrous, and wonderful." Ralph, who grew up in Lebanon, reflected each evening on his encounters with the whales that day in Mexico. "I was imagining in my head how my life would have been different had I grown up in the U.S. and been able to go to a school that offered [marine biology]." Turns out a boy from Beirut and this boy from Philadelphia shared a dream of being marine biologists and neither received much enthusiasm from their parents. Both took every opportunity to escape into the *Undersea World of Jacques Cousteau* and to the decks of the Starship *Enterprise*, yielding a wide-eyed spirit of wonder, optimism, and desire to "boldly go" that today remains as strong.

"My father was a union leader. I grew up in a house where the door is always open. There are always workers in the house and there is no hierarchy. I remember once the trash person came to collect the trash, and I said, 'Mama, the trashman is here.' She just bolted out of the kitchen and said, 'What did you call him?' I was seven years old. She continued, 'He has a name! Go ask him what his name is." And Ralph did.

"I remember walking with my father in the south of Lebanon, and an old man asked, 'May I have a dime?' I was ten years old, maybe eleven.

And I said, 'What for?' And my father grabbed me and pushed me to the ground, put his foot on my chest. And he said to me, 'Who the hell are you to ask him that? Do you know how difficult it is for an old man to ask a schmuck like you for money? Do you know how far he must have descended to ask you that? Your job is either you give him or you don't give him. You don't have the right to ask him anything else beyond that, you understand? If I ever hear you ask that ever again, you're not my son.'" Ralph explained, "That's how we were raised." Ralph's humility, imagination, and curiosity make him one of the best professional colleagues you could wish for. We are collaborating on work that we hope one day will end up in Cuba.

In 2019, Ralph and several colleagues published a landmark IMF paper, "Nature's Solution to Climate Change." On its cover, a stunning graphic of a whale. Recalling his encounters with the blue whales and evening debriefs with the whale-watching group, his thoughts turned to climate change and carbon. "I had seen this 100-foot wise creature, and she and I knew that she had more carbon than all of us together. That's how it all started." With the blue whale, Ralph took carbon sequestration beyond plants and into the world of wildlife. "You take the whale from a victim to basically turn it around so it is a key factor in the fight against climate change. So why not help save the whales, then the whale will help save us. So that's how we wrote the article."

Based on carbon prices in 2019—which have since mushroomed—Ralph's team calculated the value of a whale to be $2 million based on carbon alone. But a carbon-packed whale carcass sinking to the bottom of the sea and keeping that carbon out of the atmosphere for decades is only part of the story. Humans share much in common with whales: We are both intelligent mammals, give live birth to our young, maintain close family units, etc. Most important in this context, both of our species poop. But when a whale poops, it can be magic for the climate.

Upwelling of cold, nutrient-rich water from the depths is key to the survival of phytoplankton, the microscopic algae that drift with the current and form the base of the marine food chain. Upwelling zones exist around the world where unseen undersea rivers collide with continents. Much of the water has nowhere to go but up and the cold waters from thousands of feet below deliver nutrients to the surface. As phytoplankton thrive, zooplankton—the larvae of crustaceans, fish, and many other species—eagerly feed upon the massive concentrated blooms. It's a grand-scale opera of predator and prey—the very foundation of life in the oceans—happening right in front of our eyes, yet virtually unnoticed. But in the open ocean, typically devoid of nutrients, phytoplankton doesn't have enough nitrogen and phosphorous to support its growth. Those regions, the majority of our oceans, are referred to as "ocean deserts," vast expanses of crystal blue waters devoid of the nearshore plankton growth that colors the water green, awash with sediments and nutrients from the land. But whale poop is chock-full of the nutrients that phytoplankton need to thrive . . . and thrive they do. When you add up all the poops of a whale over its lifetime and the number of whales a-poopin', the numbers start to turn heads. But now consider that, in the case of blue whales, the global pre-whaling populations are estimated at 350,000. Today's population is 2 to 4 percent of that. It's not hard to extrapolate back a few hundred years and realize that all that whale poop translated into a significant amount of carbon sequestration. Knowledge of this so-called whale pump was well documented before Ralph's paper, but its incorporation into Ralph's economic models represents a major turning point.

Ralph reflects on his Mexico whale watching trip. "So at night I'm downloading these articles on my iPad as everybody's asleep. And I'm reading; I'm thinking, holy moly, look at this carbon! I created a spreadsheet and I'm like, oh my God, man, that whale is keeping 33 tons of carbon dioxide from the atmosphere! Now, how much carbon is a tree? So I googled the tree and it's 48 pounds. So it's about 1,500 trees. Lord have mercy!"

Considering more of the ecosystem—the role animals have in sequestering carbon—is part of what has set Ralph's work apart. But perhaps more important is how Ralph, drawing from his work with investment models at the IMF, has been able to turn the traditional economic model for offsetting carbon emissions on its head. The present model began to take hold in the early nineties. Back then, I was hired by the Electric Power Research Institute to explore offsetting carbon emissions from U.S. power plants in Siberian forests. In other words, protecting Siberia's massive forests would serve to sequester carbon from the atmosphere, offsetting the carbon spewing from oil-, natural gas–, and coal-powered power plants in the United States. The model was based on the value of carbon, and at the time, placing a value on carbon was very much a burgeoning science. The price of carbon was quite low back then. I worked with the Russian Academy of Sciences and even visited Russian timber companies. The latter grew more and more frustrated with me as our meeting went on. Finally, one of the company owners—clearly more money-seeking Russian mafia than arborist—exclaimed, "David! Exactly what part of the tree do you want to buy?" They didn't get it.

Fast-forward to the present day, and the basic model of carbon offsets has not changed significantly. When corporations seeking to lower their carbon emissions can't reach their targets within the confines of their facilities, they may pay to offset their emissions in various projects around the world. "Blue carbon," the carbon associated primarily with mangroves, seagrasses, and marine sediments has increasingly become recognized as a significant source of carbon sequestration and is a growing source of carbon offsets.

When making air reservations, consumers often have the opportunity to check a box and pay an additional fee to offset the carbon emissions of their journey. The science of calculating carbon sequestration has become robust and far more accurate over the years, but the economic part of the equation has barely evolved. Ralph—living in IMF's world of investment models—has brought this thinking to the carbon market. Rather than carbon offsets being a cost to corporations seeking to offset their emissions, what if carbon sequestration becomes an investment, one that can provide a cash return? Fueled by unprecedented increases in carbon prices on the world market, Ralph's vision is a revolutionary concept that could dramatically change the way we approach conservation.

"It's not only about helping nature; [the whale] is a viable product that can bring revenue for everybody involved. So basically, put in the language of economics, you're combining sustainability of conservation with economic development. I'm hearing the scientists say the oceans are dying and bemoaning the fact that no one is listening, and conservation is falling on deaf ears. And, of course, in my mind, I knew the first answer to that, which comes from economics: Anything that smells of a public good is always underfunded. Then the proverbial lightbulb went off. "I can create value for the whale services if you like." Ralph points out that the easier sell is one that responds to selfish interests. "I said, that's my crowd. My crowd is the tight-asses. What's in it for me? Why should I care? I'm living in Nebraska, I'll never see a whale in my life. What the hell do I care?" Now, with the added dimension of carbon, Ralph can say, "All of you care because that creature is saving you." But that's only the first step. The next step takes "save the whales" into new territory. The IMF thinks in terms of markets and investments. "All the finance guys are saying, 'I get it, I get it, man.' I'm speaking their language. So the next thing is, can we make a market out of it?" Potential investors said yes. Ralph uses the airline example. "You can check a box and pay five dollars, offset your carbon emissions. But that's not an investment, that's a cost. So this is an incredibly important distinguishing factor. I'm looking at this and thinking, how do I take this to the market?" Unlike selling a commodity like oil, a depletable resource, "you're selling the services of the whale." Ralph is fond of the example that he works for IMF, but IMF doesn't own him—they pay him a fee for his services. He began to build an economic investment model for carbon so that paying for whales to sequester carbon wasn't a cost—it was an investment, one that could pay a return to investors and create an income stream for local communities who can play an important role in protecting those resources.

So different was this new perspective and approach that it wasn't long before there was a knock at the door. It was a herd of elephants. Working in Gabon, Ralph points out that elephants are responsible for a 7 percent increase in carbon sequestration (by felling trees and promoting new growth of trees; dead trees—known to scientists as "mortmass"—release carbon very slowly over decades, and restoring elephant populations can drive that number much higher). The government of Gabon is on board with this new model—they'll have a new source of income, and perhaps best of all, the local communities will also have a new source of revenue, now having strong economic incentive to protect their elephants, and that's in addition to the elephants' value for ecotourism. With benefits accruing in the local economies, the scales tip away from poaching.

Ralph's team, my former Johns Hopkins graduate students, and the team at Rookery Bay National Estuarine Research Reserve are working to bring this model to the Western Everglades ecosystem, near Naples, Florida. The dream of Rookery Bay's superintendent, Keith Laakkonen, is to have restoration of Rookery Bay pay for itself. But we hope we can do better than that. Using Ralph's team's model, investors will receive a return on their investment and local communities and the State of Florida

213

will receive a portion of the proceeds, all beneficiaries of a growing carbon economy.

Finally, the other profoundly important distinguishing factor in Ralph's model is that investments are not short-lived. The model is designed so that investments are committed to protect the ecosystem in perpetuity. "So you look after the ecosystem and you will enjoy nature in perpetuity, you and all of generations to come. And it will bring you wealth in perpetuity," Ralph exclaims placing the coda on this emerging strategy that has world-changing potential.

Our discussion moved to Cuba, and Ralph's eyes lit up. "Cuba is an amazing opportunity. Can you imagine? My message [to Cuba] is this: You are sitting on an incredible source of wealth. And the potential is huge. And it comes from looking after your nature, not killing your nature. So you look after it, you will enjoy nature in perpetuity, you and all of generations to come. It will bring you wealth beyond in perpetuity." I'm sure it won't be long before Ralph accompanies me to visit with our Cuban colleagues.

The oceans, of course, are vast, and coral reefs spread throughout the world. Could Ralph's new economic approach work under such circumstances? Nancy Knowlton observes, "Essentially, all coral reefs are within the Exclusive Economic Zone (EEZ) of individual countries, because they're shallow-water ecosystems. So they're nearshore. So in many cases, like in Australia, they have the capability to essentially pay for whatever reef work they want to do and the same would be true of the United States. Now, of course, a lot of coral reefs live on the shorelines of developing countries, and those countries don't have the same kinds of economic resources." In Ralph's work, those costs can become investments, and that includes for Cuba, which may otherwise not be in a position to fully protect or restore its coral reef ecosystems.

Game Face

The demise of coral reefs has deeply affected many of us scientists and conservationists. The very thing we cherished, that set our careers on their paths, is disappearing. Some have quietly given up, while others still wrestle with whether corals can be saved. We conservationists are often admonished if we get too negative. We need to leave people with hope.

We need to have our game face on and engender that hope. Without it, all is lost. Nancy Knowlton gets that. She founded the Earth Optimism Campaign at the Smithsonian Institution. And she knows how to inspire while still being true to the facts. "For the situation with coral reefs. what we're trying to do is keep all the pieces in place. It's probably going to get worse before it gets better. But if we act to manage those local stressors and act aggressively about climate change, reefs will be around and they will recover. I mean, marine ecosystems in general, and reefs are no exception. They have an amazing ability to bounce back once the stresses have been relieved, and so I think that's true for reefs as well." I'm with Nancy and grateful for her optimism—it is sweet therapy and a reminder that our imagination, innovation, and ability to learn from our mistakes have given us an arsenal of tools that together give us a fighting chance to restore coral reefs. We know it requires scientists, data, and even dead things in jars. It also requires economists, social scientists, and policy experts. It will require strong, sustained, unified, and enlightened political leadership. And it will take collaboration that transcends political borders and political differences.

CHAPTER NINETEEN
SALTWATER DIPLOMATS

The soft power of science has the potential to reshape global diplomacy.

—Ahmed Zewail

The Blueprint

With the bright Cuban flag once again flying in Washington, DC, and about to enter the Cuban embassy as the outdoor reopening ceremony came to a close, I was approached by a member of the Cuban diplomatic corps I did not recognize. He apparently recognized me. "If not for your work, we wouldn't be here today." It was a lovely compliment I would hear from several Cuban officials that day, referring, of course, not to me personally, but to the small but dedicated group of U.S. NGOs who invested years to build strong scientific programs, and along the way, friendship and trust. Josefina Vidal, head of U.S. relations, delivered the same message as we embraced at the embassy entrance. Knowledge of the relationships we had been able to build, and the successes of our collaborative work, had trickled into the diplomatic dialogue, tangible illustrations of what Cubans and Americans working together could accomplish, and in doing so, lay a foundation of enduring trust and friendship. Somewhere along our journey we had become accidental diplomats—saltwater diplomats; it was marine scientists who quietly and steadfastly built strong relationships between our countries where official diplomats and politicians had for decades fallen short. For the first time in years, there were positive, inspiring examples of Cubans and Americans collaborating toward the common good, a vision for a possible future of friendship.

It's fair to say that our venture into diplomacy wasn't purely accidental. In 2014, I took first secretary of the Cuban Interests Section Warnel Lores to lunch in Washington, DC. He and Cuban ambassador José R. Cabañas, PhD, had long supported our collaboration, and we were exploring ways to convey the importance of this work and the compelling reasons that our governments should be working together on joint marine science and conservation efforts. We exchanged a number of ideas until Warnel outstretched his arms, palms up, "Why not a meeting with your Congress?" He smiled as he awaited my reaction, knowing it was a long shot and perhaps expecting me to roll my eyes and tell him it was simply impossible. But I didn't—the idea had merit. It could help provide a new perspective to members of Congress, convey the importance and urgency for improving collaboration, and, selfishly, make it easier for NGOs like ours to work in Cuba. We shook hands and I pondered how in the world we might make this happen.

The event came together faster than I could have imagined, thanks to Senator Whitehouse and his staff. A congressional briefing was scheduled in the Capitol at 9 a.m. on May 8, 2014, to include both Republicans and Democrats, members of the House and Senate, including Representatives Barbara Lee, Charles Rangel, José Serrano, Jim McGovern, Sam Farr, and Chris Van Hollen, along with Senators Jeff Flake, Sherrod Brown, Patrick Leahy, and, of course, Sheldon Whitehouse. Briefing the committee would be myself, Ambassador Cabañas, Robert Muse, and three Cuban scientists who would travel to DC from Havana: Drs. Jorge Angulo, Julio Baisre, and Fabián Pina. I also invited Dr. Robert Heuter, director of the Center for Shark Research at Mote Marine Laboratory in Sarasota, Florida. I had brought Bob to Cuba for the first time and he took the opportunity to establish a very successful shark research program with Cuban scientists. Politically speaking, I thought his work made a great example of the need for international collaboration. Many sharks travel great distances, crossing from Cuban waters into U.S. waters and vice versa, a compelling case for international collaboration. Representatives from the State Department, NOAA, and other agencies were invited to observe.

As was too often the case, getting visas for three Cubans to come to the United States was both a sprint and a marathon. Even for a congressional briefing, the process dragged. I called the State Department. I called the U.S. Interests Section in Havana. I called Senator Whitehouse's office. But Jorge, Julio, and Fabián had heard nothing. I had booked their flights on an American Airlines charter to Miami and a connecting flight that would get them to DC the evening before. The clock was ticking. They would not only need to go to the U.S. Interests Section to pick up their visas, but they would have to be interviewed before their visa was approved. In addition, Fabián would be traveling six hours by car from Cayo Coco to Havana. On the day before the flight, they still didn't have their visas. The next morning, I reached Julio at home. He had just returned from the U.S. Interests Section where he was turned away. There was no visa. He had resignation in his voice, clearly crushed by the outcome. I frantically called the State Department and was surprised to receive a callback within the hour: Julio's visa was ready! I called him at home again. "Julio, you have to go back to the Interests Section! Your visa is ready!" Time was short. "You can make it," I pleaded, "you'll make your flight!" Later I was relieved to hear from Julio that Fabián was already at the airport, visa in hand. But sadly, Jorge would not make it.

It was then in the hands of the American Airlines charter, and that did not go so well. As was too common in those days for the unreliable charter flights between Cuba and the United States, the flight was delayed by an hour. Then two. Then four. They arrived in Miami to a near-empty airport, well after the last flight to DC. I rebooked them on the first flight in the morning, but it would land at 9:20 a.m., after the hearing was scheduled to begin. The briefing was off for now. We went ahead with a planned public event that afternoon in collaboration with the Center for International Policy and we were joined by David A. Balton, deputy assistant secretary for Oceans and Fisheries at the State Department. While, without diplomatic relations, the conversation was limited, it was fruitful and conveyed an enthusiasm to find ways to work together—government to government. After the event, I heard from Senator Whitehouse's office. We'd have a scaled-down version of the briefing in his office.

When Julio, Fabián, and Robert Muse and I arrived, we were joined by Ambassador Cabañas, Josefina Vidal, David Balton, and representatives of NOAA and other agencies. Staffers from Senator Leahy's office joined us, and Senator Whitehouse opened the meeting. It was clear that we were speaking to an enthusiastic audience. After an overview of our work together, I fired up a brief PowerPoint presentation to propose three goals: First, to commit to joint marine research activities between our two countries covering fish, coral reefs, climate, etc. Second, to establish an "International Peace Park," borrowing the term from the area in the Red Sea established by Israel and Jordan, a body of water where joint research could be encouraged while serving as a symbol of cooperation between the two countries. At its essence, the proposal called for an MPA that might be jointly managed, or a network of MPAs, connected by currents and biology. And third, I proposed efforts by both countries to facilitate marine research, to make it easier for us to work together—easier to export equipment and, especially, easier to get visas! With no inkling that the big announcement from Presidents Castro and Obama to restore diplomatic relations was just seven months away, we had, during that meeting, managed to create a rough blueprint of what would be the first memoranda of understanding between Cuba and the United States, all of which were focused on environmental collaboration, especially marine science and conservation. And among the memoranda was something akin to an International Peace Park. NOAA, the National Park Service, and Cuba's National Center for Protected Areas agreed to work together on "establishing sister-sanctuary relationships between Guanahacabibes and Banco de San Antonio in Cuba, and Florida Keys and Flower Garden Banks national marine sanctuaries in the United States—recognizing that these places are all inextricably linked through the flow of the ocean." Banco de San Antonio, off Cuba's northwestern coast, is the area we had first explored aboard the *Boca del Toro* a decade earlier. Back then, the area had united two groups of scientists. Today, it was uniting two nations.

The Scrap of Paper

Years earlier, after the Bush administration tightened restrictions on Cuba and all but eliminated academic programs, many of us came to the inescapable conclusion that we would need to join together to advance collaborative academic research between the United States and Cuba and push back on academic restrictions. Further, because of the political situation, there was limited data sharing and participation by Cuba in in-

THE REMARKABLE REEFS OF CUBA

ternational forums. We needed a collaborative or coalition as a vehicle to facilitate data sharing to provide a better understanding of the science and conservation of the Gulf of Mexico and Western Caribbean and elevate collaboration to a new level.

Toward that end, Wayne Smith of the Center for International Policy, Wes Tunnell of the Harte Research Institute, Robert Muse, and I came together to create with CITMA what ultimately would be called the "Trinational Initiative for Research and Conservation in the Gulf of Mexico and Western Caribbean." The initiative would also include Mexico. I would lead the initiative through its first five years. Of course, the United States was represented only by NGOs—U.S. governmental agencies were unable to participate given the lack of diplomatic relations and political climate. Excited about our inaugural meeting—to be held in Cancún—I opened the first day of our three-day meeting and welcomed the group of around 50 in attendance. As we went around the table for introductions, the meeting disintegrated into a Cuban-led political diatribe about the U.S. embargo against Cuba. Try as I did, it was simply impossible to get the group back on track. The Mexican delegation could only sit and watch. By late afternoon and adjournment for the day, we were left with virtually no consensus nor progress. I grew desperate. If ever saltwater diplomacy was needed, it was now.

As the group was enjoying cocktails, I pulled the leader of the Cuban delegation, Jorge L. Fernández Chamero, aside to work with him quietly, privately and delicately. I pleaded with him to provide me with what the Cubans thought should be the priority issues for the initiative to address. I emphasized how important it was to have this information before we convened in the morning, lest we run adrift for a second day.

The next morning, I sipped my coffee and tapped on my laptop, trying to formulate the words I would use to open the meeting that might set us back on course, but I was coming up empty. Participants were beginning to take their seats and I felt their eyes on me at the front of the room as I fumbled with the microphone. It was then that I saw Chamero approach. He discreetly handed me a small scrap of paper without saying a word and took his seat at the table. I unfolded the paper and smiled. Written on the paper were six priority areas for joint research identified by the Cubans, from sea turtle conservation to MPAs. Within a couple of hours, with support from the U.S. and Mexican participants, those six priorities formed the core of successful collaboration by the Trinational Initiative and still do to this day.

Generations

I was caught off guard one November afternoon years ago when a student I did not know approached me at Havana's Palacio de Convenciones during a major conference on marine science in Cuba. He shook my hand and thanked me for our support of the expeditions that allowed him to do his master's research. At a meeting at CIM later that week, I was struck by the fact that I was the oldest one at the table. The next generation that we had helped along was now in charge of the institute and had blossomed into young scientists thinking of Americans as friends and colleagues, not enemies.

The next generation of scientists is tasked with solving today's unprecedented, grand-scale problems. The spectacle of the Cuyahoga River on fire seems quaint and easily remedied compared to the scale and complexity of rescuing our coral reefs and addressing the global threats they face. Today's generation must work across disciplines like never before—biologists, chemists, climatologist, economists, physicists, etc. It means we must change our perspective and the way we see the world and its myriad connections. And, of course, it means not only working across disciplines but also working across borders. Cuba and the United States have a beautiful neighborhood to care for. And neither of us can do it alone. From where I sit, the next generation is more than up to the task, already carrying important collaboration to new levels.

In countless ways, the island of Cuba is unique. And when it comes to coral reefs, Cuba is again, unique. Here an island of thriving corals flourishes amid a world of corals dying and disappearing. In this mysterious corner of the world where time seems to have stopped, I find hope. Hope that the rich ecosystems of this beautiful island will endure. And hope that Cuba's coral reefs continue to share their tantalizing secrets, secrets that can offer clues to protecting and restoring coral reefs elsewhere, including a special place I still remember in the Florida Keys, just 90 miles to the north.

EPILOGUE

*The most beautiful people we have known are those who have known defeat,
known suffering, known struggle, known loss, and have found their way out
of those depths.*

—Elisabeth Kübler-Ross

The bulk of this book was written during Cuba's most difficult
times since the Special Period of the 1990s. Frequently it appeared that the writing would be overtaken by current events, like
the COVID-19 pandemic; the faltering Cuban economy; the breakdown
of Cuba's health-care system; the July 11, 2021, protests in Cuba; the free
fall of the Cuban peso and estimates by economists of an inflation rate of
300–500 percent as of early 2022; and proposed U.S. legislation against
the Cuban government that could be the harshest since the Helms-
Burton Act of the mid-nineties.

Tightened restrictions by the Trump administration—perpectuated
under Biden—have limited U.S. visitation and choked remittances from
families in the United States to their relatives in Cuba. Then came the
pandemic. Tourism was halted and the tourism-dependent economy went
into free fall. Shortages of foods and medicines ensued. At the time of this
writing, my friend Liuvys Nuñez, trying to support her two children, has
found prices highly inflated on the black market. She finds this period
worse than what she remembers of the Special Period. Her sister Lis
Nuñez, our consultant in Cuba and close friend, struggles to rehabilitate,

feed, and care for 50 cats in a refuge she built on a Havana rooftop. Her dedication and resourcefulness astound me. She recently informed me that a dozen eggs now costs $10, half an official month's salary.

On the Isle of Youth, Keily Ibañez Castellanos, an ecotourism professional and close friend I have worked with, has been preparing and selling meals to support her mother and son. With no tourists, she has no work. For some time, travel from the island to Havana was not permitted because of the pandemic. So she and her family have been isolated. She also finds the situation worse than the Special Period. When her mother fell gravely ill, there was no medicine available for her. Her mother's condition was deteriorating, and Keily feared the worst. Thanks to the generosity of a close friend who is a medical doctor in the United States, I was able to obtain the medicine she needed and sent it to a relative of Keily in Miami. After passing what was probably 10 sets of hands, the medicine miraculously made it to the Isle of Youth, where her mother has since made a full recovery. But now she says there is neither milk nor eggs to feed her family.

Some feel that the current situation has not reached the depths of the Special Period because of the internet. Aside from the ability to communicate, resourceful young Cubans are finding novel ways to make a living online. Mandao, a smartphone app, is the Cuban equivalent of Uber Eats, emerging during the pandemic lockdown when home delivery of food has been in high demand. Other Cubans are earning money through cryptocurrency and a new generation of blockchain games. In other words, play games and win cryptocurrency. It's a thing. Cuban friends of mine have made as much as $300 per month, and they tell me others have made much more. The internet and Cubans' access to it has made this book possible. Unable to visit Cuba for more than two years because of the pandemic, I was not able to interview a number of Cuban colleagues I had hoped to. But thanks to WhatsApp and Telegram, I was able to conduct interviews with many others.

During nearly two years of lockdown, social media has been an integral part of Cuban life. The protests of July 11, 2021—the largest in many years—resulted in a government crackdown and interruption of internet services, but not before social media spread images around the world of protesters being beaten by authorities. "Social networks are playing a very important role in the democratization of the society. It's happening.

That's one of the reasons why they were so afraid with the demonstrations on the street and they shut down the internet," Jesús Noguera told me. The Cuban government's actions sparked outrage, and support for the embargo has become more entrenched. There are ongoing efforts to further crack down on Cuba, including the "DEMOCRACIA Act," introduced in October 2021, which would choke investment in Cuba by international corporations, something that could be economically catastrophic for Cuba.

For years, explanations of the health of Cuba's coral reefs were based mostly on speculation. So I was delighted during the writing of this book to learn that much of what had been speculation is now supported by hard data. With universities and other institutions closed for so long, it's been next to impossible for much research to continue during the pandemic. New data will have to wait for some time, and Cuba's reefs will continue to cling to their mysteries.

It has become a tradition during my trips to Cuba to speak briefly in Cuban grade schools. Of course, I teach them about coral reefs. It's always a surprise to them—and many other Cubans—when I tell them that their country has some of the healthiest coral reefs left on the planet. I tell them, *"Necesitan ustedes sentirese orgullosos!* (You need to feel proud!)" And they do. And today those young students and their families should also feel especially proud that in a world of unrelenting hardship, Cuban people and Cuban corals share in common one of life's most important qualities: *Resiliencia.*

ACKNOWLEDGMENTS

This book covers more than two decades, making it an impossible feat to adequately express my gratitude to all who deserve it. My family has been unwavering in their enthusiasm and support from the earliest days when the book was just an idea, through its journey to the finish line. A heartfelt thank-you to cousins Catherine and David Behrend, who became my teammates and supporters on this trek. My brother Alan, my wife Svetlana, and my daughter Anna gave me strength and confidence at every step

This book begins at Seacamp, a marine science camp in the Florida Keys where I set my course in life. More than a marine education institute, Seacamp is a family of enduring and extraordinary friendships that I treasure. Special thanks to its late founder and director, Irene Hooper, and her sister and current director, Grace Upshaw, who carries the torch and spirit forward. My deepest thanks to Dr. James A. Bohnsack who was my mentor at Seacamp and who inspired me to pursue my career in marine science and conservation, and whom I was delighted to interview for this book. Other Seacampers interviewed whom I thank include Robert P. Beech and Dr. Don R. Levitan. And thanks to Donna Coffin, whose bloodcurdling scream will ever linger in my memories and is a key part of this book.

To the warm and dedicated Cubans who welcomed me with open arms 20 years ago. Like Seacamp, so many I have worked with over the years are family, too. Most of my work has been focused at the Center for Marine Research at the University of Havana (CIM), and I have been privileged to work alongside both faculty and students. Special thanks to the late Dra. María Elena Ibarra, CIM director; Dr. Jorge Angulo, who followed Dra. Ibarra as director; and Dra. Patricia González, its current director. Sincere thanks to former CIM vice directors Dr. Gaspar González Sansón and Dra. Ana María Suarez Alfonso. Warmest thanks to Dra. Consuelo María Aguilar Betancourt, Dr. Rogelio Díaz-Fernández, Dra. Julia Azana-Ricardo, and many others. They are integral parts of this story and my life. Our gratitude to Comandante de la Revolución Guillermo García Frías who lent us his support—and his boat—to conduct our research.

Much of this story would have been impossible without friend and colleague Dr. Julio A. Baisre, whose vision and advocacy led to the creation of one of the world's largest and healthiest marine reserves, Gardens of the Queen. Cuban ambassador José R. Cabañas, PhD, and first secretary of the Cuban Interests Section Warnel Lores shared the vision of using a positive example of Cuba–U.S. collaboration, our joint work in marine science and conservation, to help improve collaboration between the two countries. Their vision worked and helped create a foundation of trust that aided in the eventual restoration of diplomatic relations, an effort that would have been impossible without the work of Josefina Vidal of the Ministry of International Relations. The ongoing support of Ambassador Cabañas and Josefina Vidal was essential to the success of our work. Sincere thanks to Dra. Soraya M. Castro Mariño at Centro de Investigaciones de Política Internacional (CIPI), the Center for International Policy Research, for reaching out to and including scientists and conservationists in international policy dialogues. I am forever grateful for the friendship, guidance, and humor of Dr. Sergio Pastrana, foreign secretary and executive director of the Cuban Academy of Sciences.

My thanks to Reinaldo "Nene" Borrego Hernandez and the remarkable community of Cocodrilo on the Isle of Youth. Our work together has been important and inspiring. It is Dr. Fabián Pina of Cuba's Center for Ecosystem Research (CIEC) who brought us to the magnificent Gardens of the Queen, a place to which we would return again and again to draw hope from an extraordinarily healthy coral reef ecosystem and learn how protection of large areas of the ocean can make an enormous difference. Dr. Pina leads important research on the area that is a significant part of this account. His wife, Tamara Figueredo Martín, also at CIEC, is one of Cuba's few environmental economists, a key partner as our work broadened to focus on environmental economics. Divemasters Noel López (the "shark whisperer"), Antonio "Tony" Luis Cardenas, and Andrés Jiménez helped us understand Gardens of the Queens and graciously welcomed and educated the hundreds of American visitors Ocean Doctor brought to Cuba. Tony's important work with lionfish has been enlightening. Many thanks to Dra. Lisandra Torres Hechavarría, associate dean at the University of Havana, Department of Tourism, for her friendship and important insight on tourism in Cuba.

Lis Nuñez, Ocean Doctor's representative in Cuba, a dear friend who has sacrificed everything to build a refuge and rehabilitation "center" for abandoned kittens and young cats, has been such an important guide to help me understand daily life in Cuba. I'm still learning. Thanks also to her sister, Liuvys. Jesús Noguera Ravelo, who runs the travel service CU-BACAREO, has been an exceptional guide for many of my groups. He has also been an exceptional friend. He, too, has helped me understand Cuban life. Close friend Daylin Muñoz Nuñez, whom I knew in Cuba and now in the United States where she resides, has been a scientific colleague and has helped me understand Cuban culture and the challenges faced by the next generation of Cuban scientists. Cuban American Vilma Albelay, and her family, including her father Ramón, have also been my guides—and my friends. My dear friend and colleague Keily Ibañez Castellanos on the Isle of Youth has helped me understand that unusual island and the struggles she and her fellow islanders have faced during the pandemic. Thanks to Mayra Alonso of Marazul Charters, who has helped me achieve the impossible: travel to Cuba more than 100 times.

I am deeply grateful to those I interviewed for this book not mentioned above. Many graciously guest lectured for my class on marine stewardship and conservation at Johns Hopkins University (where my students deserve thanks for their great questions, and some of their class work is part of this book). The wisdom and experience of Drs. Nancy Knowlton, Sant Chair for Marine Science at Smithsonian Institution, and Jeremy Jackson, emeritus professor at Scripps Institution of Oceanography and senior scientist emeritus at Smithsonian Institution, are present throughout this book. Their understanding of coral reef ecosystems is unparalleled. I was incredibly fortunate to cochair the Everglades Coalition with my friend and now–principal deputy assistant secretary for Fish and Wildlife and Parks, Shannon Estenoz. Excerpts of her guest lecture to my students about the Everglades were key in illustrating the inextricable link between land and water. Margarita Fernández, executive director of the Cuba Agroecology Institute in Vermont, and Dr. Paul Bierman of the University of Vermont provided critical insight about agricultural practices in Cuba. John Hocevar, oceans campaign director at Greenpeace and fellow submersible pilot, sheds light (literally) on the importance of the lesser-known corals, deepwater corals, that are at risk like their tropical cousins. He continues to lead efforts to protect them. Ellen Rugh, director of research and programs at the Center for Responsible Travel, helped my students understand how travel impacts the Caribbean and other regions, both environmentally and socioeconomically. Jeff Orlowski provided important insights into his remarkable film, *Chasing Coral*, and its role in increasing awareness of the devastating impacts of climate change on coral reefs. Robert L. Muse, Esq., the leading expert on legal and regulatory issues between the United States and Cuba, has helped me wrap my head (at least partially) around the maddeningly complex rat's nest of regulations surrounding the U.S. economic embargo against Cuba. Our work in Cuba would have been impossible without his guidance. Wayne Smith, former head of the U.S. Interests Section in Havana and an outspoken leader on U.S.–Cuba policy, has been a constant inspiration and source of perspective on the state of relations between our countries. Many thanks to Nick Drayton, whom I consider one of my closest friends, and Dr. Stephanie Wear of the Nature Conservancy shared war stories and insight on what it took to establish a marine protected area in the U.S. Virgin Islands. Dr. Ralph Chami, the "whale whisperer," is assistant director of the Institute for Capacity Development, Western Hemisphere Division, at the International Monetary Fund. He has turned conservation on its head, and his work and spirit are part of the optimistic future presented in the book. I am delighted to be working with him on exciting new projects. Jean-Michel Cousteau, son of Jacques-Yves Cousteau and head of Ocean Futures, shared his accounts from the 1985 Cousteau expedition to Cuba. I'm still grateful to him and his organization for their support during the BP oil spill. Senator Sheldon Whitehouse knows marine biology better than some marine biologists. An ocean leader, he founded the Senate Oceans Caucus. We have spent time together in Cuba—he has been a leader on Cuba policy as well as ocean policy. He is a friend and has my deepest respect. And my students loved him.

Special thanks to Robert F. Kennedy Jr. who traveled with me to Cuba where he met with Fidel Castro. From his meeting, we learned just how much of an influence Jacques-Yves Cousteau was on Castro in setting the future environmental path of Cuba. My sincere thanks to the *60 Minutes/* CBS team, including producer Andy Court, associate producer Anya Bourg, correspondent Anderson Cooper, and producer and Havana-based CBS news chief producer Portia Siegelbaum. They masterfully told the important story of coral reefs at a time when no one else was.

My sincere thanks to the Ocean Doctor team, including Mary Kadzielski and Ximena Escovar-Fadul, for their incredible work. Intern Rachael Hughen wrote much of our 2018 report on unsustainable travel discussed in this book, which was also coauthored by independent consultant Melissa Mooney Walton and Elizabeth Newhouse of the Center for International Policy. My sincere thanks also for the support and guidance of the Ocean Doctor board, including Bob Frank, Adam Ravetch, and renowned ocean conservation leader Dr. Sylvia Earle, with whom I have been privileged to collaborate for more than 20 years and have shared many an adventure at sea—and under the sea, of course. Shari Sant Plummer has been a strong supporter of our work from the very beginning and has accompanied me to Cuba many times. She heads Code Blue Foundation, and I am grateful for the foundation's support of our Cuba work. We are also grateful to many other foundations that have supported us, including the Ford Foundation, the Baum Foundation, Summit Foundation, Charlotte's Web Foundation, and the Bay and Paul Foundations, not to mention the hundreds of individual donors whose support has made a profound difference. Thanks also to the other U.S. and international NGOs and academic institutions working in Cuba—together, with our Cuban colleagues, we are making a difference.

Though I have authored many articles, reports, and blogs over the years, writing a book is nothing less than an expedition. I am grateful to my agent, John Willig, for his guidance and ever-present support, and to Jake Bonar at Globe Pequot–Prometheus Books who immediately understood the story I wanted to tell and supported me every step of the way. Thanks also to talented fiction writer Tiffany Butler, who helped guide me and provide badly needed moral support as I entered the alien world of publishing a book. Tiffany recommended that I read Courtney Maum's *Before and After the Book Deal*, which was pure therapy. Trudy Hale's hospitality at her writers' retreat provided the solitude that I so needed to be able to transform pages of writing into a book, with only the sound of an occasional freight train to remind me to take a break. Finally, it's a scientific fact that writing is impossible without coffee. My caffeinated thanks to Philz Coffee and the LINE Hotel in Washington, DC, for a place to plug in my laptop, and for the caffeine buzz, of course.

BIBLIOGRAPHY

AGRRA. "Coral Disease Outbreak," 2019. https://www.agrra.org/coral-disease -outbreak/.

"All-Inclusive: An Evolved Model Takes Center Stage. FocusOn Report—Americas." JLL Research—Hotels & Hospitality, 2017.

Allen, Irwin. *The Sea Around Us* (Film), 1953.

Asamblea Nacional de Cuba. "Law No. 81—Environmental Law." (1997). https:// digitalrepository.unm.edu/la_energy_policies/VALUE THE FUTURE.

Bak, R. P. M., and M. J. E. Carpay. "Densities of the Sea Urchin *Diadema Antillarum* before and after Mass Mortalities on the Coral Reefs of Curaqao," n.d., 4.

"Ballast Water | National Invasive Species Information Center." Accessed May 3, 2020. https://www.invasivespeciesinfo.gov/subject/ballast-water.

Bariche, Michel, et al. "Genetics Reveal the Identity and Origin of the Lionfish Invasion in the Mediterranean Sea." *Scientific Reports* 7, no. 1 (December 2017): 6782. https://doi.org/10.1038/s41598-017-07326-1.

Bark, Amberly, et al. "Stormwater Management and the Chesapeake Bay": White Paper. Ocean Stewardship & Conservation, Johns Hopkins University, Advanced Academic Programs, 2020.

Beck, Gregory, Robert Miller, and John Ebersole. "Mass Mortality and Slow Recovery of *Diadema Antillarum*: Could Compromised Immunity Be a Factor?" *Marine Biology* 161, no. 5 (May 1, 2014): 1001–13. https://doi.org/10.1007/ s00227-013-2382-6.

Beyer, Hawthorne L., et al. "Risk-sensitive Planning for Conserving Coral Reefs under Rapid Climate Change." *Conservation Letters* 11, no. 6 (November 2018). https://doi.org/10.1111/conl.12587.

Bierman, Paul, et al. "¡Cuba! River Water Chemistry Reveals Rapid Chemical Weathering, the Echo of Uplift, and the Promise of More Sustainable Agriculture." *GSA Today* 30, no. 3 (March 2020): 4–10. https://doi.org/10.1130/GSATG419A.1.

Birkeland, Charles. *Life and Death of Coral Reefs*. New York: Chapman & Hall, 1999.

"Bleaching of Coral Reefs in the Caribbean: Hearing Before a Subcommittee of the Committee on Appropriations, United States Senate, One Hundredth Congress, First Session: Special Hearing." Washington, DC: U.S. Government Printing Office, 1988. https://books.google.com/books/about/Bleaching_of_coral_reefs_in_the_Caribbea.html?id=BpgQAAAAIAAJ.

Bohnsack, James A., Laura Jay W. Grove, and Joseph E. Serafy. "Use of the Term Harvest When Referring to Wild Stock Exploitation." *Fishery Bulletin* 119, no. 1 (December 28, 2020): 1–2. https://doi.org/10.7755/FB.119.1.1.

Borunda, Alejandra. "Ocean Acidification." *National Geographic*, August 7, 2019. https://www.nationalgeographic.com/environment/oceans/critical-issues-ocean-acidification/.

Briggs, John C. "An International Symposium: The Sea-Level Canal Controversy." *Defenders of Wildlife News*, January 1973, 60–62.

Burke, Lauretta, et al. "Reefs at Risk Revisited." Washington, DC: World Resources Institute, 2011.

Callwood, Karlisa A. "Examining the Development of a Parrotfish Fishery in The Bahamas: Social Considerations & Management Implications." *Global Ecology and Conservation* 28 (August 2021): e01677. https://doi.org/10.1016/j.gecco.2021.e01677.

Castro Speech Data Base—Latin American Network Information Center, LANIC. Accessed November 11, 2021. http://lanic.utexas.edu/project/castro/db/1966/19660910.html.

Center for Biological Diversity. "Lawsuit Launched Over DDT Ocean Dumping off Southern California." Accessed July 6, 2021. https://biologicaldiversity.org/w/news/press-releases/lawsuit-launched-over-ddt-ocean-dumping-off-southern-california-2021-05-27/.

Chami, Ralph, Thomas Cosimano, Connel Fullenkamp, and Sena Oztosun. "A Strategy to Protect Whales Can Limit Greenhouse Gases and Global Warming." *Finance & Development—International Monetary Fund*, December 2019, 34–38.

Chesher, Richard H. "Transport of Marine Plankton Through the Panama Canal." *Limnology and Oceanography* 13, no. 2 (April 1968): 387–88. https://doi.org/10.4319/lo.1968.13.2.0387.

"The Clean Water Act at 40: There's Still Much Left to Do." Yale University. Accessed June 9, 2021. https://e360.yale.edu/features/the_clean_water_act _at_40_theres_still_much_left_to_do.

"Clean Water Act Dramatically Cut Pollution in U.S. Waterways | Berkeley News." Accessed June 9, 2021. https://news.berkeley.edu/2018/10/08/clean -water-act-dramatically-cut-pollution-in-u-s-waterways/.

"Coastal Capital: Economic Valuation of Coastal Ecosystems in the Caribbean | World Resources Institute." Accessed May 25, 2018. http://www.wri.org/ our-work/project/coastal-capital-economic-valuation-coastal-ecosystems -caribbean.

Columbus, Christopher. *The Four Voyages: Being His Own Log-Book, Letters and Dispatches with Connecting Narratives.* Translated by J. M. Cohen. Revised ed. edition. Harmondsworth, England, UK: Penguin Classics, 1992.

Committee on Interventions to Increase the Resilience of Coral Reefs, Ocean Studies Board, Board on Life Sciences, Division on Earth and Life Studies, and National Academies of Sciences, Engineering, and Medicine. A Research Review of Interventions to Increase the Persistence and Resilience of Coral Reefs. Washington, DC: National Academies Press, 2019. https:// doi.org/10.17226/25279.

"Coral Bleaching Events, AIMS." Accessed October 8, 2021. https://www.aims .gov.au/docs/research/climate-change/coral-bleaching/bleaching-events.html.

Cornu, Jean-Paul. "Cuba: Waters of Destiny." *Cousteau's Rediscovery of the World.* Cuba: Turner Broadcasting Network, 1986.

Cowen, R. K., C. B. Paris, and A. Srinivasan. "Scaling of Connectivity in Marine Populations." *Science* 311, no. 5760 (January 27, 2006): 522–27. https://doi .org/10.1126/science.1122039.

Cramer, Katie L, et al. "Prehistorical and Historical Declines in Caribbean Coral Reef Accretion Rates Driven by Loss of Parrotfish." *Nature Communications* 8 (January 23, 2017): 1–8. https://doi.org/10.1038/ncomms14160.

"Cuba Puts a Hold on Licensing Some Private Businesses | Insights | Greenberg Traurig LLP." Accessed May 23, 2018. https://www.gtlaw.com/en/ insights/2017/8/cuba-suspends-licensing-private-businesses.

"Cuban Food Output Stagnates, May Decline in 2017." Reuters, October 17, 2017. https://www.reuters.com/article/us-cuba-agriculture/cuban-food-ou tput-stagnates-may-decline-in-2017-idUSKBN1CM1Z5.

Depraetere, Valerie. "African Dust Chokes Caribbean Reefs." *Nature*, November 20, 2000. news001123-4. https://doi.org/10.1038/news001123-4.

Devine, Jon. "Clean Water Act at 45: Despite Success, It's Under Attack." NRDC, October 18, 2017. https://www.nrdc.org/experts/jon-devine/clean -water-act-45-despite-success-its-under-attack.

Dominguez, Jorge I. "What You Might Not Know About the Cuban Economy." *Harvard Business Review*, August 17, 2015. https://hbr.org/2015/08/what -you-might-not-know-about-the-cuban-economy.

"The Gardens of the Queen." *60 Minutes*. New York, NY: CBS, December 18, 2011. Producers: Andy Court, Anya Bourg, Portia Siegelbaum. Correspondent: Anderson Cooper.

Gonzalez, G., et al. "Present Condition of Coral Reefs and Associated Ecosystems in the Northwest Region of Cuba." 11th International Coral Reef Symposium. Ft. Lauderdale, Florida, 2008.

Gorbachev, Mikhail. "Mr. Bush, Tear Down This Wall." *Washington Post*, September 1, 2005, Opinion section.

Haman, Dorota Z., and Mark Svendsen. "Managing the Florida Everglades: Changing Values, Changing Policies." *Irrigation and Drainage Systems* 20, nos. 2–3 (August 2006): 283–302. https://doi.org/10.1007/s10795-006 -9008-9.

Hendren, M. J., et al. "Climate Change Resilience Using Green Infrastructure and the Potential of Carbon Valuation to Offset Restoration Cost in Southwest Florida." White Paper. Ocean Stewardship & Conservation, Johns Hopkins University, Advanced Academic Programs. 2020.

Hubbard, Dennis K. "The Demise of Acropora Palmata in Caribbean Reefs: Past, Present and Future." Joint annual meeting of American Society of Agronomy (ASA), Crop Science of America (CSA), and Soil Science of America (SSSA), Caribbean Reefs Session, Houston, Texas, October 5–9, 2008.

Humboldt, Alexander von. *Equinoctial Regions of America*, volume 2, n.p.: Outlook Verlag, 1859. https://www.google.com/books/edition/Equinoctial_Re gions_of_America/.

"The IUCN Red List of Threatened Species." IUCN, 2021. https://www.iuc nredlist.org.

Jackson, J. B. C. "Reefs Since Columbus." *Coral Reefs* 16 (December 1, 1997): S23–32. https://doi.org/10.1007/s003380050238.

Jackson, J. B. C., et al. "Historical Overfishing and the Recent Collapse of Coastal Ecosystems." *Science* 293, no. 27 (July 2001): 629–37.

Jackson J. B. C., et al. (editors). (2014) *Status and Trends of Caribbean Coral Reefs: 1970–2012*. Global Coral Reef Monitoring Network, IUCN, Gland, Switzerland.

Jackson, Jeremy, and Ayana Elizabeth Johnson. "We Can Save the Caribbean's Coral Reefs." *New York Times*, September 19, 2014.

"Jardines de la Reina National Park, Cuba." Accessed May 25, 2018. http://glo balconservation.org/projects/jardines-de-la-reina-national-park-cuba/.

Johnson, Darlene R., Nicholas A. Funicelli, and James A. Bohnsack. "Effectiveness of an Existing Estuarine No-Take Fish Sanctuary within the Kennedy Space Center, Florida." *North American Journal of Fisheries Management* 19, no. 2 (May 1999): 436–53.

Kam, Yulang, et al. "The Iraqi Crab, *Elamenopsis Kempi* in the Panama Canal: Distribution, Abundance and Interactions with the Exotic North American Crab, *Rhithropanopeus Harrisii.*" *Aquatic Invasions* 6, no. 3 (September 2011): 339–45. https://doi.org/10.3391/ai.2011.6.3.10.

Kaye, Ken. "Nation's Meanest Hurricane Devastated the Keys 80 Years Ago." sun-sentinel.com. Accessed August 25, 2019. https://www.sun-sentinel.com/local/broward/fl-labor-day-hurricane-anniversary-20150827-story.html.

Keiser, David A., and Joseph S. Shapiro. "Consequences of the Clean Water Act and the Demand for Water Quality." *Quarterly Journal of Economics* 134, no. 1 (February 1, 2019): 349–96. https://doi.org/10.1093/qje/qjy019.

Kleinhaus, Karine, Ali Al-Sawalmih, Daniel J. Barshis, Amatzia Genin, Lola N. Grace, Ove Hoegh-Guldberg, Yossi Loya, et al. "Science, Diplomacy, and the Red Sea's Unique Coral Reef: It's Time for Action." *Frontiers in Marine Science* 7 (2020): 90. https://doi.org/10.3389/fmars.2020.00090.

Knowlton, Nancy. "Sea Urchin Recovery from Mass Mortality: New Hope for Caribbean Coral Reefs?" *Proceedings of the National Academy of Sciences* 98, no. 9 (April 24, 2001): 4822–24. https://doi.org/10.1073/pnas.091107198.

———. *Citizens of the Sea: Wondrous Creatures From the Census of Marine Life.* 8/16/10 edition. Washington, D.C: National Geographic, 2010.

Knowlton, Nancy, and Jeremy B. C. Jackson. "Shifting Baselines, Local Impacts, and Global Change on Coral Reefs." *PLoS Biology* 6, no. 2 (February 26, 2008): e54. https://doi.org/10.1371/journal.pbio.0060054.

Kough, A., R. Claro, K. C. Lindeman, and C. B. Paris. "Decadal Analysis of Larval Connectivity from Cuban Snapper (Lutjanidae) Spawning Aggregations Based on Biophysical Modeling." *Marine Ecology Progress Series* 550 (May 25, 2016): 175–90. https://doi.org/10.3354/meps11714.

Lafferty, Kevin D., James W. Porter, and Susan E. Ford. "Are Diseases Increasing in the Ocean?" *Annual Review of Ecology, Evolution, and Systematics* 35, no. 1 (December 15, 2004): 31–54. https://doi.org/10.1146/annurev.ecolsys.35.021103.105704.

Lessios, H. A. "The Great *Diadema Antillarum* Die-Off: 30 Years Later." *Annual Review of Marine Science* 8, no. 1 (January 3, 2016): 267–83. https://doi.org/10.1146/annurev-marine-122414-033857.

Lessios, H. A., D. R. Robertson, and J. D. Cubit. "Spread of *Diadema* Mass Mortality Through the Caribbean." *Science, New Series* 226, no. 4672 (October 19, 1984): 335–37.

Levitan, Don R., Peter J. Edmunds, and Keeha E. Levitan. "What Makes a Species Common? No Evidence of Density-Dependent Recruitment or Mortality of the Sea Urchin *Diadema Antillarum* after the 1983–1984 Mass Mortality." *Oecologia* 175, no. 1 (May 2014): 117–28. https://doi .org/10.1007/s00442-013-2871-9.

Levitan, D. R., W. Boudreau, J. Jara, and N. Knowlton. "Long-Term Reduced Spawning in *Orbicella* Coral Species Due to Temperature Stress." *Marine Ecology Progress Series* 515 (November 18, 2014): 1–10. https://doi .org/10.3354/meps11063.

Lindeman, Ken, et al. (2001). "Transport of larvae originating in southwest Cuba and the Dry Tortugas: Evidence for partial retention in grunts and snappers." Proceedings of the Gulf and Caribbean Fisheries Institute, 52, 732–747.

Marshall, Michael. "Why We Find It Difficult to Recognise a Crisis." Accessed May 7, 2020. https://www.bbc.com/future/article/20200409-why-we-find -it-difficult-to-recognise-a-crisis.

Mega, Emiliano Rodriguez. "Cuba Adds Climate to Its Constitution." *Nature* 567 (March 14, 2019): 155.

Menendez, Mario. "Mexican Magazine Has Exclusive with Fidel." *Sucesos*, September 10, 1966.

"Million-Dollar Reef Sharks | Pew." Accessed June 9, 2018. http://pew .org/2yISm3S.

"The Mississippi/Atchafalaya River Basin (MARB)." Overviews and Factsheets. US EPA, March 24, 2015. https://www.epa.gov/ms-htf/mississippiatchafa laya-river-basin-marb.

Mort, Terry. *The Hemingway Patrols: Ernest Hemingway and His Hunt for U-Boats*, annotated edition. Scribner, 2009.

Muse, Robert L. "The President Has the Constitutional Power to Unilaterally Terminate the Embargo on Cuba." Global Americans (blog), October 8, 2020. https://theglobalamericans.org/2020/10/the-president-has-the-con stitutional-power-to-unilaterally-terminate-the-embargo-on-cuba/.

Nature. "Cuba: The Accidental Eden | The Causeway to Cayo Coco | *Nature* | PBS," September 23, 2010. http://www.pbs.org/wnet/nature/cuba-the -accidental-eden-the-causeway-to-cayo-coco/5808/.

NBC News. "Cuba Sees a Decline in Tourism as U.S. Travel to the Island Drops Sharply." Accessed January 23, 2022. https://www.nbcnews.com/news/ latino/sharp-decline-u-s-travel-cuba-spurs-overall-drop-tourism-n868856.

"NOAA National Ocean Service Education: Corals." Accessed May 23, 2018. https://oceanservice.noaa.gov/education/kits/corals/coral12_references.html.

Oceana Europe. "The Importance of Sharks." Accessed June 28, 2021. https://europe.oceana.org/en/importance-sharks-0.

"OP Notice to Shipping No. N-1-2008: Vessel Requirements." Panama City, Panama: Autoridad del Canal de Panama (ACP), January 1, 2008.

Orlowsi, Jeff. *Chasing Coral.* Netflix, 2018. https://www.pbs.org/hemingwayadventure/cuba.html.

Panama Canal, Pub. L. No. 92–96, § Committee on Merchant Marine and Fisheries. Subcommittee on Coast Guard, Coast and Geodetic Survey and Navigation (1971).

Phinney, Jonathan T., et al. "Using Remote Sensing to Reassess the Mass Mortality of *Diadema Antillarum* 1983–1984." *Conservation Biology* 15, no. 4 (August 3, 2001): 885–91. https://doi.org/10.1046/j.1523-1739.2001.015004885.x.

Pina Amargós, Fabián, Tamara Figueredo-Martín, and Natalia A Rossi. "The Ecology of Cuba's Jardines de la Reina: A Review." *Revista de Investigaciones Marinas* 41, no. 1 (June 2021): 16–57.

"Presidential-Proclamation-No.-5030-EEZ.Pdf," March 10, 1983.

"Pro Poor Tourism—Definition." Accessed May 26, 2018. http://www.ecotourdirectory.com/pro-poor-tourism.htm.

Puig, Pere, et al. "Ploughing the Deep Sea Floor." *Nature* 489, no. 7415 (September 1, 2012): 286–89. https://doi.org/10.1038/nature11410.

Raloff, Janet. "Wanted: Reef Cleaners." *Science News* 160, no. 8 (August 25, 2001): 120. https://doi.org/10.2307/4012705.

"Researchers Find Coral Reefs at Risk When Sharks Overfished | Earth | EarthSky," September 27, 2013. https://earthsky.org/earth/researchers-find-coral-reefs-at-risk-when-sharks-overfished/.

"Restoring the Everglades, An American Legacy Act. Report of the Committee on Environment and Public Works, United States Senate." Washington, DC: United States Government Printing Office, July 27, 2000.

Reyes-Bonilla, Héctor, and Eric Jordán-Dahlgren. "Caribbean Coral Reefs: Past, Present, and Insights into the Future." In *Marine Animal Forests: The Ecology of Benthic Biodiversity Hotspots*, edited by Sergio Rossi, Lorenzo Bramanti, Andrea Gori, and Covadonga Orejas, 31–72. Cham: Springer International Publishing, 2017. https://doi.org/10.1007/978-3-319-21012-4_2.

Ritz, Leah T., and Gabriel Jácome (advisor). "The Current State of Populations of *Diadema Antillarum* on Isla Colón in Bocas Del Toro, Panamá, 25 Years after Mass Mortality," n.d., 26.

Roberts, C. M. "Effects of Marine Reserves on Adjacent Fisheries." *Science* 294, no. 5548 (November 30, 2001): 1920–23. https://doi.org/10.1126/sci ence.294.5548.1920.

Rocliffe, S., and R. H. Leeney, "Research Briefing: Bottom Trawling and the Climate Crisis." London, UK: Blue Ventures, 2021.

Roff, George, et al. "The Ecological Role of Sharks on Coral Reefs." *Trends in Ecology & Evolution* 31, no. 5 (May 2016): 395–407. https://doi .org/10.1016/j.tree.2016.02.014.

Rothchild, John. *Up for Grabs: A Trip Through Time and Space in the Sunshine State.* Gainesville: University of Florida Press, 2000.

Rubinoff, Ira. "The Sea-Level Canal Controversy." *Biological Conservation* 3, no. 1 (October 1970): 482–85.

Ruppert, Jonathan L. W., et al. "Caught in the Middle: Combined Impacts of Shark Removal and Coral Loss on the Fish Communities of Coral Reefs." *PLoS ONE* 8, no. 9 (September 18, 2013): e74648. https://doi.org/10.1371/ journal.pone.0074648.

Sala, Enric, et al. "Protecting the Global Ocean for Biodiversity, Food and Climate." *Nature* 592, no. 7854 (April 1, 2021): 397–402. https://doi .org/10.1038/s41586-021-03371-z.

Salgado, Jorge, et al. "The Panama Canal after a Century of Human Impacts." Preprint. *Ecology*, September 23, 2019. https://doi.org/10.1101/777938.

Schill, Steven R., et al. "Site Selection for Coral Reef Restoration Using Airborne Imaging Spectroscopy." *Frontiers in Marine Science* 8 (2021): 1022. https:// doi.org/10.3389/fmars.2021.698004.

Shinn, Eugene A., et al. "African Dust and the Demise of Caribbean Coral Reefs." *Geophysical Research Letters* 27, no. 19 (October 1, 2000): 3029–32. https://doi.org/10.1029/2000GL011599.

Shore, Amanda, et al. "On a Reef Far, Far Away: Anthropogenic Impacts Fol- lowing Extreme Storms Affect Sponge Health and Bacterial Communi- ties." *Frontiers in Marine Science* 8 (2021): 305. https://doi.org/10.3389/ fmars.2021.608036.

"Silent Invasion: The Spread of Marine Invasive Species via Ships' Ballast Wa- ter." Gland, Switzerland: WWF International, 2009.

Slezak, Michael. "More Coral Bleaching Feared for Great Barrier Reef in Com- ing Months." *Guardian*, November 3, 2017, sec. Environment. http://www .theguardian.com/environment/2017/nov/03/more-coral-bleaching-feared -for-great-barrier-reef-in-coming-months.

Sutherland, Kathryn Patterson, et al. "Human Pathogen Shown to Cause Disease in the Threatened Eklhorn Coral *Acropora Palmata.*" Edited by

Steve Vollmer. *PLoS ONE* 6, no. 8 (August 17, 2011): e23468. https://doi
.org/10.1371/journal.pone.0023468.

Szmant, Alina M. "Nutrient Enrichment on Coral Reefs: Is It a Major Cause of
Coral Reef Decline?" *Estuaries* 25, no. 4 (August 1, 2002): 743–66. https://
doi.org/10.1007/BF02804903.

Thornburg, Jack. "Eco-tourism and Sustainable Community Development in
Cuba: Bringing Community Back into Development," *Journal of International and Global Studies*, vol. 9, no. 1 (2017), article 2. https://digitalcom
mons.lindenwood.edu/jigs/vol9/iss1/2.

"The Tobago Hilton Story." Accessed May 22, 2018. http://www.raymondan
dpierre.com/articles/article46.htm.

United States Geological Survey. "Reported Lionfish Sightings: Animated Map
(1985–2020)." Accessed November 11, 2021. https://www.usgs.gov/media/
images/reported-lionfish-sightings-animated-map-1985-2020.

U.S. Department of Commerce, National Oceanic and Atmospheric Administration. "What Is Coral Bleaching?" Accessed May 23, 2018. https://ocean
service.noaa.gov/facts/coral_bleach.html.

U.S. EPA, Office of Water. "Combined Sewer Overflows (CSOs)." Overviews
and Factsheets. U.S. EPA, October 13, 2015. https://www.epa.gov/npdes/
combined-sewer-overflows-csos.

"Value of Reefs | Reef Resilience Network." Accessed June 3, 2018. http://www
.reefresilience.org/coral-reefs/reefs-and-resilience/value-of-reefs/.

Vianna, G. M. S., et al. "Corrigendum to 'Socio-Economic Value and Community Benefits from Shark-Diving Tourism in Palau: A Sustainable Use
of Reef Shark Populations' [*Biological Conservation* 145 (2012) 267–277]."
Biological Conservation 156 (November 2012): 147. https://doi.org/10.1016/j
.biocon.2012.11.010.

Walton, Melissa Mooney, et al. "A Century of Unsustainable Tourism in the
Caribbean: Lessons Learned and Opportunities for Cuba." Washington,
DC: Ocean Doctor, 2018.

"Watershed | Chesapeake Bay Program." Accessed June 16, 2021. https://www
.chesapeakebay.net/discover/watershed.

"WCS Wild View: How Sharks Help Keep Coral Reefs Healthy." Accessed June
28, 2021. https://blog.wcs.org/photo/2021/02/22/how-sharks-help-keep
-coral-reefs-healthy-education-africa/.

Weisberger, Mindy. "Coral Reefs Have 'Halos,' and They Can Be Seen from
the Heavens." livescience.com, April 24, 2019. https://www.livescience
.com/65310-coral-reef-halos.html.

What Is Geotourism? | The Bahamas. Accessed January 23, 2022. https://bahamasgeotourism.com/entries/what-is-Geotourism/37e3a3b8-878a-4428-9705-baac2c1fa602.

whitehouse.gov. "Statement by the President on Cuba Policy Changes," December 17, 2014. https://obamawhitehouse.archives.gov/the-press-office/2014/12/17/statement-president-cuba-policy-changes.

Wijsman, Jeroen. "Panama Canal Extension: A Review on Salt Intrusion into Gatun Lake." Wageningen, The Netherlands: IMARES Wageningen UR, December 2013.

Wilson, Suzanne Leigh. "When Disorder Is the Order: Cuba During the Special Period." Dissertation. Berkeley: University of California, 2011.